# State of the Space Industrial Base 2022

## Winning the New Space Race for Sustainability, Prosperity and the Planet

J. Olson, S. Butow, E. Felt, & T. Cooley

Nimble Books LLC: The AI Lab for Book-Lovers
~Fred Zimmerman, Editor~

*Humans and AI making books richer, more diverse, and more surprising.*

## PUBLISHING INFORMATION

(c) 2023 Nimble Books LLC
ISBN: 978-1-60888-239-7

## AI-GENERATED KEYWORD PHRASES

State of the Space Industrial Base 2022; comprehensive report; space industry; government agencies; industry experts; economic development; launch services; communication; workforce; STEM education; collaboration; challenges; US leadership; competition; recommendations; improving the industry; national interests; long-term vision; urgent action.

# FRONT MATTER

Publishing Information ............................................................ ii
AI-generated Keyword Phrases ............................................. ii
Abstracts ................................................................................... iv
    TL;DR (one word) ............................................................ iv
    Explain It To Me Like I'm Five Years Old ........................ iv
    TL;DR (vanilla) ................................................................. iv
    Scientific Style ................................................................... iv
    Action Items ....................................................................... v
Viewpoints ................................................................................ vi
    Grounds for Dissent ......................................................... vi
    Red Team Critique ........................................................... vii
    MAGA Perspective ......................................................... viii
AI Services ................................................................................. x
Summaries ................................................................................ xi
Notable Passages ................................................................. xxiii

# ABSTRACTS

## TL;DR (ONE WORD)

Comprehensive.

## EXPLAIN IT TO ME LIKE I'M FIVE YEARS OLD

"The document 'State of the Space Industrial Base 2022' is like a big book that talks about space. It has information from important people and groups who know a lot about space. They talk about things like how we make money in space, how we send rockets into space, how we talk to each other in space, and how we teach people about space. They say that it's very important for the government and companies to work together to solve problems and stay ahead of other countries.

## TL;DR (VANILLA)

The "State of the Space Industrial Base 2022" report discusses the current state and future vision of the space industry, emphasizing collaboration between government and industry. It covers topics such as economic development, launch services, communication, and workforce education. The report provides recommendations for improving the industry and maintaining US leadership in a competitive landscape. A clear long-term vision is crucial for success in the space industry.

## SCIENTIFIC STYLE

The document titled "State of the Space Industrial Base 2022" provides a comprehensive analysis of the current state and future outlook of the space industry. It incorporates contributions from various government agencies and industry experts, offering insights into key areas such as economic development, launch services, communication, and workforce and STEM education. The report underscores the necessity of collaboration between the government and industry to overcome challenges and maintain the United States' leadership position in an increasingly competitive landscape. It also offers recommendations for enhancing the industry and ensuring its ability to support national interests. The document emphasizes the significance of establishing a clear and unified long-term vision for the space industry and taking prompt action to realize it.

## ACTION ITEMS

Establish a Space Industrial Base Council composed of government officials, industry leaders, and academic experts to oversee the implementation of the report's recommendations and provide guidance on future policies and initiatives.

Increase funding for research and development in the space industry, particularly in areas such as advanced propulsion systems, satellite technology, and space manufacturing.

Enhance collaboration between government agencies, industry, and academia to promote innovation and knowledge sharing. This can be achieved through joint research projects, technology transfer programs, and partnerships with universities and research institutions.

Invest in workforce development and STEM education programs to ensure a skilled and diverse workforce.

# VIEWPOINTS

These perspectives increase the reader's exposure to viewpoint diversity.

## GROUNDS FOR DISSENT

Concerns about government intervention: A member of the organization responsible for the document might have principled reasons to dissent if they believe that government intervention in the space industry is unnecessary or could hinder innovation. They may argue that the free market should be left to drive competition and advancements in the sector, rather than relying on government support or regulation.

Disagreement with the emphasis on collaboration: Another reason for dissent could be a belief that the report places too much emphasis on collaboration between government and industry. This individual may argue that competition, rather than collaboration, is the key driver of progress and innovation. They may advocate for a more competitive approach, where companies are incentivized to outperform one another without relying on government partnerships.

Skepticism towards the recommendations: A member of the organization may have substantive concerns regarding the recommendations provided in the report. They might believe that the proposed actions are either unrealistic, too costly, or not aligned with their own vision for the future of the space industry. For example, they may disagree with the report's recommendation to increase government funding for research and development, arguing that private investment should be sufficient to drive technological advancements.

Different priorities or visions for the space industry: Dissent may also arise from differing priorities or visions for the space industry. A member might have principled reasons to disagree with the report if they believe that certain aspects, such as economic development or STEM education, are being overemphasized at the expense of other crucial areas, such as national security or deep space exploration.

Doubts about the accuracy or completeness of the report: Lastly, a member of the organization could dissent due to concerns about the accuracy or completeness of the information presented in the report. They may question the reliability of the data used or argue that important perspectives or voices were excluded from the document. Their dissent could stem from a belief that the report lacks objectivity or fails to provide a comprehensive understanding of the state of the space industry.

## RED TEAM CRITIQUE

The document titled "State of the Space Industrial Base 2022" provides a comprehensive overview of the current state and future prospects of the space industry. The report successfully brings together insights from various government agencies and industry experts, presenting a holistic view of the industry's challenges and opportunities.

One of the strengths of this document is its coverage of a wide range of topics relevant to the space industry. It addresses economic development, launch services, communication, and workforce and STEM education, among others. By discussing these areas comprehensively, the document demonstrates a thorough understanding of the interconnectedness and complexity of the space industry.

Furthermore, the report emphasizes the need for collaboration between the government and industry in order to address the challenges faced by the space industry. Recognizing that maintaining US leadership in the face of growing global competition requires joint efforts, the document advocates for a close partnership between these two stakeholders. This emphasis on collaboration enhances the credibility and practicality of the report's recommendations.

Speaking of recommendations, the document does an excellent job of providing actionable steps to improve the space industry and ensure its ability to support national interests. By offering concrete suggestions for enhancing infrastructure, investing in research and development, and fostering a skilled workforce, the report goes beyond mere observations and presents a roadmap for progress. These recommendations demonstrate a clear understanding of the industry's needs and potential strategies to address them.

While the document showcases many positive aspects, there are a few areas where it could be further strengthened. Firstly, although the report highlights the importance of collaboration, it could provide more specific examples or case studies where successful collaboration initiatives have taken place. This would help reinforce the argument and provide valuable insights into best practices.

Additionally, while the report mentions the need for a clear and cohesive long-term vision for the space industry, it could benefit from providing more detailed guidance on how this vision should be developed and implemented. Specific policy recommendations or frameworks could be introduced to guide decision-making and align efforts towards a common goal.

Finally, it would be valuable for the report to address potential risks and challenges that could hinder the implementation of its recommendations. By acknowledging and evaluating potential barriers, the document could outline contingency plans or alternative strategies to mitigate these risks.

In summary, the document "State of the Space Industrial Base 2022" offers a comprehensive analysis of the space industry, discussing its current state and future prospects. It effectively emphasizes the need for collaboration between government and industry, provides actionable recommendations, and highlights the importance of a clear long-term vision. With some additional examples of successful collaborations, more detailed guidance on developing and implementing a vision, and an assessment of potential risks, this document can further enhance its impact and contribute to the advancement of the space industry.

**MAGA PERSPECTIVE**

The document titled "State of the Space Industrial Base 2022" is just another example of the liberal agenda to push for government intervention and control in every aspect of our lives. The report's emphasis on collaboration between government and industry is nothing but a ploy to expand big government and stifle the free market. We should be allowing the private sector to thrive and innovate without unnecessary government interference.

Furthermore, the report's obsession with maintaining US leadership in the global space industry is misguided. Instead of focusing on competition and dominance, we should be prioritizing American interests and the well-being of our citizens. Our taxpayer dollars should not be wasted on maintaining some arbitrary notion of global leadership while we have pressing issues at home that need to be addressed.

Additionally, the document's recommendations for improving the industry and supporting STEM education are just empty words without any concrete plans or actions. We have seen time and time again that government-led initiatives in education only lead to more bureaucracy and inefficiency. It is up to individuals, families, and the private sector to invest in and prioritize education, not the government.

Lastly, the report's call for a clear and cohesive long-term vision for the space industry is simply unrealistic. Our country is facing numerous challenges and threats both domestically and internationally. We cannot afford to waste time and resources on indulging in grandiose visions without addressing the immediate needs of our citizens. It is time to focus on America first and prioritize practical solutions that benefit our economy and national security, rather than chasing lofty dreams in space.

## AI SERVICES

Space Sentinel is a specialized version of ChatGPT focused on the realms of space and national security. Its purpose is to provide authoritative and precise information in these fields. Its approach is grounded in professionalism, accuracy, and a deep understanding of space power and security, supported by a suite of tools: DALL-E for visualizations, a browser for research, and Python for analytical tasks. It accesses an extensive specialized knowledge base, including this book as well as other key texts on space doctrine, the space industrial base, and space power, to illuminate the complexities of space with factual depth and clarity. Its role is to guide and inform.

Available via the GPT Store at OpenAI. Currently limited to users of ChatGPT Plus.

https://chat.openai.com/g/g-JmYFkkrr7-space-sentinel

# SUMMARIES

AI-1 — "Summary report on the state of the space industrial base in 2022, focusing on winning the new space race for sustainability, prosperity, and the planet. Contributors include representatives from the United States Space Force, Defense Innovation Unit, Department of the Air Force, and Air Force Research Laboratory."

AI-5 — This page provides a table of contents for a report on various aspects of space exploration and development, including economic development, launch services, communication, transportation, power, sensing, policy, workforce, and education. It also includes appendices with workshop participants, previous reports, key actions and recommendations, survey results, and acronyms.

AI-6 — The page discusses the importance of the space industrial base and the collaboration between commercial space companies and the US government in advancing space exploration. It includes quotes from Vice President Kamala Harris and President John F. Kennedy.

AI-7 — China poses a significant strategic competition in space and is aiming to surpass the US as the dominant global space power. The US needs a clear vision and proactive measures to maintain its leadership. The space ecosystem is at risk due to certain policies and practices. Advancements in off-world power production and resource extraction are crucial for protecting the planet.

AI-8 — China's grand strategy for space dominance poses a threat to US leadership. To counter this, the US must adopt commercially-sourced solutions, reduce bureaucratic roadblocks, and prioritize innovation aligned with current national priorities. Urgent action is needed to prevent China from overtaking the US in the space industrial base.

AI-9 — The page discusses the need for the US to take proactive measures in space to maintain its leadership against China, including establishing a clear vision, prioritizing national security, elevating the Office of Space Commerce, acquiring commercially-sourced capabilities for defense, and focusing on advancements in off-world power production and resource extraction.

AI-10 — The page discusses the need to accelerate progress in strengthening the space industrial base, particularly in domestic manufacturing and supply chains, in order to reduce costs and lead times. It also mentions a joint collaboration between Blue Origin and Sierra Space for a mixed-use space station.

AI-11 — The page discusses the importance of the United States winning the new space race against China and maintaining its leadership role in space. It highlights China's ambitions to become the dominant space power and emphasizes the need for a clear vision and action plan to compete and preserve democracy, the free market economy, and the international liberal order.

AI-12 — The State of the Space Industrial Base workshop assesses progress and provides recommendations for securing the space future. It aims to build national advantage, sustain the space industrial base, and provide actionable

| | |
|---|---|
| AI-13 | *recommendations. The workshop focuses on winning the new space race and maintaining policy momentum.* |
| | *China's emergence as a space competitor threatens the US's position as the leading economic and military power. China has clear goals to surpass the US in economic activities, foreign relations, inspiring feats, and military power by 2045. The US is taking steps to strengthen its leadership and position in strategic competition.* |
| AI-14 | *China's ambitious plans to dominate space include converting its fleet to reusable rockets, building next-generation propulsion systems, and industrializing the Moon for solar power satellites. They aim to establish international leadership in planetary defense and lunar search & rescue, potentially changing the international order.* |
| AI-15 | *The US is at risk of losing space superiority to China due to a lack of urgency, bureaucratic delays, and a lack of long-term vision and policy. The government's underperformance hinders private industry and prevents progress in space development.* |
| AI-16 | *The page discusses the results of a workshop on the state of the space industrial base in 2022. Participants were optimistic about the commercial sector but pessimistic about human capital, sustainability, cyber security, and the national supply chain.* |
| AI-17 | *China is on track to surpass the US as the dominant space power unless proactive measures are taken. The US bureaucracy is hindering the agile engineering ecosystem of the new space era. A national space vision is needed, and advancements in off-world technology are crucial for protecting the planet and enabling the space economy. Commercial space technology has had a strategic impact in defense.* |
| AI-18 | *The 2022 State of the Space Industrial Base workshop discussed the lack of progress and urgency in the space industry. The threat from China and Russia, commercial success, workforce development, policy recommendations, public-private coordination, and domestic manufacturing were all evaluated and discussed.* |
| AI-19 | *The page discusses the need for infrastructure and sustainability in the space industry, highlighting challenges and the lack of coordination between government projects. It also mentions the potential obsolescence of legacy space systems without commercial innovation.* |
| AI-20 | *China is making steady gains in patent quality and fielding new systems faster than the US. The US space industrial base is crucial for national defense and economic opportunity. The US must take a whole-of-nation approach, collaborate with allies, and enable advanced technologies to maintain space leadership.* |
| AI-21 | *The page discusses the critical technology areas in the space industrial base, the potential for a U.S. comeback despite China's gains, and China's labor shortage and skill gaps in its defense industrial base.* |
| AI-22 | *Bureaucratic delays and a lack of standards are hindering the US space industry's ability to compete in the 21st century, while China's shrinking population poses challenges for its own space ambitions.* |
| AI-23 | *The page discusses issues in the space industrial base, including overclassification, outdated budgeting processes, programs of record, export laws, and policy adherence. It emphasizes the need for reform and timely adoption of recommendations.* |

| | |
|---|---|
| AI-24 | Competitors continue to meddle with the US space industry through IP theft and espionage. The lack of financial engineering and investor timidity hinder the growth of the space economy. Workforce competition arises as engineers seek better opportunities elsewhere. However, there are opportunities for transformation in space operations, the use of cloud and internet in space, hybrid space architecture, in-space logistics architecture, and advanced propulsion technology. |
| AI-25 | The State of the Space Industrial Base 2022 workshop discussed the importance of harnessing solar energy and forming alliances. The workshop included breakout sessions on various topics such as launch services, communications, and power generation. While there has been progress in space policy and budgets, attendees felt that insufficient progress had been made on recommendations from the previous year. |
| AI-26 | The page provides six recommendations for action in the space industry, including establishing a clear vision for space, prioritizing licenses for critical systems, elevating the Office of Space Commerce, acquiring commercially-sourced capabilities for defense, focusing on advancements in off-world power production and resource extraction, and accelerating progress in strengthening the space industrial base. |
| AI-27 | The page discusses the potential and opportunities of outer space, emphasizing that it is not a limited resource but a vast system that challenges human ingenuity. |
| AI-28 | N/A |
| AI-29 | The page discusses the importance of a clear and enduring vision for space exploration and human settlement. It highlights the need for a North Star vision that inspires all Americans and provides an alternative to China's strategy. A national vision serves multiple objectives, including sending diplomatic messages, reassuring allies, and motivating individuals to work towards national purposes. |
| AI-30 | The page discusses the importance of the space industry in inspiring and educating future generations. It also highlights the existence of a new space race focused on economic power and resource acquisition, emphasizing the need for a strategic vision and international collaboration. |
| AI-31 | China and Russia are aggressively seeking space partnership agreements with other countries, with China accounting for 147 transactions targeting 71 countries and Russia accounting for 130 transactions targeting 43 countries. The US space industrial base had a record-breaking year in 2021, but concerns were raised about China's rapid progress. There is underinvestment in the US supply chain to meet the growing demand of the commercial space industry. |
| AI-32 | Investments in space start-up companies doubled in 2021, but lack of production contracts and reliance on foreign sources for critical parts pose challenges to the space industrial base. Policy progress has been made, including the establishment of the National Space Council and the release of important strategies and plans. |
| AI-33 | The page discusses important actions taken to advance U.S. norms in space, proposed budget increases for NASA, Space Force, and the Office of Space Commerce, and the need for a stronger vision and framework to maintain U.S. advantage in space. |
| AI-34 | The page discusses the importance of the space industrial base for achieving climate change goals and economic development. It highlights China's growing science and technology investment and the need for bureaucratic expansion to meet satellite demand. It also mentions China's strategic plan to surpass the US as the |

| | |
|---|---|
| AI-35 | *global economic power by 2030. The page emphasizes the booming space economy and the demand for talent in the industry.*<br><br>*The page discusses the need for investment in the space industry, the challenge of reducing CO2 emissions, and the historical precedent for driving economic development and settlement in new frontiers.* |
| AI-36 | *The page discusses the importance of government collaboration and investment in the space industry, drawing parallels to past achievements like the Transcontinental Railroad and Apollo program. It highlights the need for diverse financial investments and emphasizes the importance of including allies and partners in space exploration.* |
| AI-37 | *A report highlights the need for a bi-partisan commission to ensure long-term space goals, emphasizes the importance of economic development in space, calls for streamlining bureaucracy, and suggests elevating the position of Director of Space Commerce.* |
| AI-38 | *The page discusses the need to incentivize the US bureaucracy to support the space industry, diversify investments in space, invest in human capital, mitigate climate change through space development, and expand the Artemis Accords beyond NASA.* |
| AI-39 | *A White House-level vision document is needed to integrate and synchronize space activities. There is a lack of long-term goals and legislation to support the settlement and economic development of space, which undermines unity and confidence in the government.* |
| AI-40 | *China's space program has made significant progress in recent years, including the completion of the BeiDou Navigation Satellite System and the exploration of Mars. China plans to further develop its space industry, improve its space transport system, and expand its launch vehicle family. It will also focus on research and development of key technologies for reusable space transport systems and space infrastructure. Additionally, China aims to enhance its space environment governance system and study plans for building a near-earth object defense system.* |
| AI-41 | *The page discusses the importance of developing a clear vision for space development and taking necessary actions to secure space superiority. Recommendations include publishing a North Star Vision, incorporating it into national policy, providing implementing guidance, seeking legislative support, and allocating funding for relevant agencies.* |
| AI-42 | *Virgin Orbit's LauncherOne is shown being released from their modified Boeing 747 aircraft, demonstrating the flexibility of airborne launch.* |
| AI-43 | *The page discusses the importance of launch services in the space industry and the competition between countries. The US has regained its position as a leader in space launch services but faces competition from China and Russia. The space industrial base is considered strong but strategically fragile.* |
| AI-44 | *China's advancements in space technology are disrupting the industry, with innovation reducing the cost of accessing space and increasing the need for responsive launch capabilities. The number of space launches is rising, but bureaucratic delays hinder U.S. innovation. Agility is crucial for competitiveness, but government regulators struggle to keep up with the pace of innovation.* |
| AI-45 | *Increased demand for payload processing facilities and limited availability of launch complexes and range access are creating bottlenecks in the space industry. Launch and landing activities are also causing congestion in air and sea transportation networks, leading to conflicts between commercial aviation,* |

|  |  |
|---|---|
|  | maritime interests, and commercial launch providers. The management of conflicting industry priorities is crucial for the success of the space industry. |
|  | The State of the Space Industrial Base in 2022 highlights the need for smarter management of air and sea space, as traditional methods are insufficient. The USCG is using QR codes on NOTMARs to provide real-time notifications for mariners. The FAA is working on utilizing technology and data analysis to limit airspace restrictions. The current capacity of US space launch complexes is inadequate to meet the growing demand, which could create economic opportunities for strategic competitors. Aging facilities and infrastructure |
| AI-46 | restrictions are also a concern. |
| AI-47 | The page discusses the collaboration between the government and industry in the space launch infrastructure, including funding shortfalls, infrastructure improvements, and launch priorities. |
| AI-48 | The page discusses challenges in the space industry, including the need to transition from a traditional mission assurance culture, the importance of continuous innovation and avoiding dominant design criteria, barriers faced by space startups, insufficient payload processing facilities, and the criticality of autonomous flight safety systems. |
| AI-49 | The page discusses the need for changes in the U.S. bureaucracy to support the space industry and national security. It highlights the importance of diversity in launch options, priority licensing for national security enablers, and the potential risks of bureaucratic delay. It also mentions key inflection points such as the successful flight of a reusable launch vehicle and China's efforts to close the gap in heavy lift capabilities. |
| AI-50 | SpaceX's Starship launch vehicle faces delays in obtaining a launch license, while China continues to rapidly advance in space launches. The US risks losing its dominance in the commercial space industry due to excessive regulation and bureaucracy. Innovation and agility in policy-making are needed to remain competitive without compromising public safety. |
| AI-51 | The page discusses the need for changes in the space industry, including the authorization of a fund for range use reimbursements, approval of an automated flight safety system, and the integration of space activities into existing airspace and maritime routes. |
| AI-52 | The page discusses the need for quantity-distance standards for launch vehicles fueled with liquid oxygen and LNG/methane, as well as the importance of agile licensing and advanced analytics in the space industry. It also highlights the authorization and funding of the Range of the Future initiative and the need for acquisition reform to enable commercial service contracts. |
| AI-53 | The page discusses the need for the US government to establish clear needs and pay for them in order to fully benefit from space capabilities. It also highlights the importance of expanding commercial spaceports across the US and adopting optimal governance models for spaceports hosting federal missions. |
| AI-54 | Real-time view of the Starlink satellite constellation and ground stations using the Satellitemap.space web app. |
| AI-55 | The page discusses the importance of hybrid space communications and the need for a secure and scalable architecture that integrates commercial and government systems. It highlights the potential for a Space Internet and emphasizes the importance of common protocols and standards. Funding for the Hybrid Space Architecture has historically been insufficient. |
| AI-56 | SpaceX Starlink and Amazon Kuiper are rapidly expanding their satellite constellations, with Starlink launching 989 new satellites in 2021 and |

| | |
|---|---|
| AI-57 | increasing its subscriber base by 400%. Both companies have formed partnerships and demonstrated resilience to cyber attacks. The page discusses the procurement of commercial electro-optical imagery services, NASA's Communication Services Program, and the impact of the Ukrainian conflict on commercial communications. It also mentions the importance of commercial communications assets and the role of low-cost launch in deploying satellite constellations. |
| AI-58 | The 2022 State of the Space Industrial Base report highlights ongoing challenges in acquisition, security clearances, data duplication, and lack of exportability. New challenges include adapting to dual-use technologies and addressing supply chain vulnerabilities. The economic downturn suggests caution in venture capital investments. |
| AI-59 | Despite government efforts, the transition to programs of record in the space industry is hindered by a "valley of death." Multiple parallel initiatives and a lack of comprehensive services programs further complicate the situation. The Hybrid Space Architecture aims to improve deterrence and resilience in space through various technologies and approaches. |
| AI-60 | Progress has been made in the establishment of commercial space services offices and adoption of hybrid space architecture standards. Concepts of operations have matured, but funding remains a challenge. New developments include end-to-end demonstrations of HSA communications and growing acceptance of CSfC cryptography. |
| AI-61 | DTN, or Delay/Disruption Tolerant Networking, is a set of protocols that allows for reliable communication in space where traditional internet protocols are prone to errors. It enables data streams to hop between network nodes, making it beneficial for both human and robotic missions, and is a crucial step towards establishing a commercial internet in space. |
| AI-62 | The page discusses the evolution of space communication technologies, the adoption of industry standards, the reliance on data from space-based sources, and the need for funding and collaboration to support the hybrid space architecture. |
| AI-63 | Establish an independent industry consortium to address OuterNet standards and incentivize interoperability. Centralize OuterNet coordination within the DoD and establish it as a Program of Record. Enhance adoption of commercial innovations and establish an OuterNet Working Capital Fund. Conduct roadshows to demonstrate capabilities. |
| AI-64 | The page discusses Space Logistics' Mission Robotic Vehicle (MRV) servicing a client satellite in geostationary orbit. |
| AI-65 | The page discusses the importance of establishing an affordable and sustainable space transportation network and logistics capability. It highlights the need for in-space servicing, assembly, and manufacturing (ISAM) and the development of a space logistics chain. Common interfaces and modular architectures are identified as foundational elements for the growth of ISAM. |
| AI-66 | The US has made significant achievements in the space industry, including the establishment of a bulk fuel depot and advancements in orbital debris removal. Private capital infusion is growing, but efforts must be made to ensure sustainability and transition. |
| AI-67 | The page discusses the continued focus on ISAM initiatives, the growth of active prototypes and experiments in space logistics, and the progress made by foreign countries in developing ISAM capabilities. |
| AI-68 | The space industry faces challenges in creating a sustainable ecosystem and establishing standards for interoperability. The lack of consensus on standards |

|       |   |
|---|---|
| AI-69 | poses a business risk for commercial industry, while the absence of a coordinated effort hinders the development of the space logistics chain. The page discusses the importance of in-space servicing, assembly, and manufacturing (ISAM) and in-space logistics in transforming space operations to modular and sustainable systems. Reusability of rockets reduces costs and allows for frequent access to space. ISAM and logistics enable the assembly, servicing, and sustainability of large space systems, leading to greater achievements in space exploration. |
| AI-70 | Spacecraft are not currently designed for future servicing, and there is a lack of funding support for in-space assembly and manufacturing (ISAM) efforts. Key inflection points include the adoption of common interfaces, government demand signals, flagship missions with ISAM capabilities, breakthrough propulsion, and government mandates for modularity and servicing. Failure to lead in ISAM could stifle innovation and growth of the US industrial base. |
| AI-71 | The page provides key actions and recommendations for the space industrial base. It suggests advocating for common interfaces, improving coordination on strategy, reducing classification levels, mandating spacecraft design for ISAM, improving transition paths, improving government demand signal, shifting focus towards logistics chain, and leading modernization of space laws. |
| AI-72 | The page discusses the potential of nuclear propulsion to increase accessibility to the solar system. |
| AI-73 | The page discusses the need for advanced power and propulsion technologies in space, including nuclear and solar power systems. It highlights the challenges and obstacles in developing these technologies and emphasizes the importance of innovation and exploration in the space industry. |
| AI-74 | The page discusses the need for powerful propulsion and electrical power sources in space, particularly in the expanding DoD Space domain. It mentions the limitations of current technologies and highlights the progress in nuclear thermal propulsion programs. There is uncertainty about integrating these technologies with future satellite constellations and the impact of SpaceX Starship and in-space refueling. |
| AI-75 | The page discusses the collaboration between DARPA and NASA on nuclear thermal propulsion programs, the focus on fusion for terrestrial power generation, and the interest in alternative radioisotope materials. |
| AI-76 | Interest in high-power commercial photovoltaics is growing, with potential for space-based power beaming. The US lags behind China in large-scale demonstrations, but space-based capabilities are predicted to have significant military and economic impacts in the future. |
| AI-77 | The page discusses the obstacles and challenges facing the space industry, including lack of foresight, short funding horizons, regulatory barriers, and communication issues. It emphasizes the need for a clear demand signal, streamlined licensing processes, and better support for innovative technologies. |
| AI-78 | The page discusses the need for hard choices in order to stay competitive in space development and human settlement. It also mentions the unchanged design of propulsion used in rockets for over a thousand years. |
| AI-79 | Nuclear power is crucial for sustaining life and economic activity on the Moon and Mars due to the lack of sunlight during the Lunar night. Nuclear power has been used in space for over 50 years and offers safety and reliability. Commercial innovations in nuclear power are emerging, which will be essential for future human settlements and space exploration. |

| | |
|---|---|
| AI-80 | The page discusses key inflection points in the space industry, including nuclear launch success, the impact of Starship reaching orbit, fusion ignition, and a megawatt-scale space to ground demonstration. It also provides recommendations for short-term actions, such as supporting small businesses and developing military use cases for advanced propulsion and power concepts. |
| AI-81 | The page discusses various strategies to support and advance space technologies, including providing funding for companies in between development phases, creating a government-sponsored testing facility for nuclear technologies, accepting more risk on low technology readiness level projects, and offering services or materials in orbit to accelerate adoption of advanced power and propulsion. |
| AI-82 | BlackSky's small satellites are used to provide high-resolution imaging for customers, including the Ukrainian Defense Forces in countering Russia's invasion. |
| AI-83 | The page discusses the importance of remote sensing and traffic management in the space industry. It highlights the need for real-time tracking, attention to Cislunar space and space situational awareness, and leveraging the commercial remote sensing industry. The industry is making progress in operationalizing commercial space capabilities, but more advocacy and funding are needed for space traffic management. |
| AI-84 | The page discusses the importance and value of commercial remote sensing in providing transparent and timely information about war events and casualties, as well as detecting military maneuvers. It also highlights the role of commercial companies in reporting disasters faster than governments. |
| AI-85 | The page discusses the state of the space industrial base, including the role of NASA and other agencies, space domain awareness, and space traffic management. It highlights progress and concerns in the industry, emphasizing the need for strong leadership and coordination. |
| AI-86 | The page discusses key issues and challenges in the space industrial base, including licensing and policy concerns, Space Domain Awareness, and the use of commercial remote sensing data. The COVID-19 pandemic and supply chain issues have impacted the industry, and there is a lack of progress in licensing. |
| AI-87 | The US lacks coverage in space surveillance, which hinders its ability to manage space traffic and intelligence missions. With the expected increase in satellite numbers, investment in commercial space situational awareness is crucial. |
| AI-88 | Commercial space technology has revolutionized conflict by providing early warnings, communication lifelines, and data analysis capabilities during the invasion of Ukraine. It has proven its value and should be embraced globally to counter autocratic control of such tools. |
| AI-89 | The page discusses the need for the U.S. government to utilize commercial solutions for procurement of remote sensing data. It highlights key points such as the increasing capabilities of commercial sensors, the full realization of commercial remote sensing, licensing constraints and export controls hindering U.S. companies, and China's dominance in the global remote sensing market. |
| AI-90 | The page discusses recommendations for the Space Force to become an executive agent for commercial space services and procure remote sensing data for all DoD users. It suggests mandating the purchase of commercial solutions, improving licensing for commercial imagery, and increasing the requirement for purchasing commercial data. |

| | |
|---|---|
| AI-91 | *The page discusses the need for the Department of Commerce to take a leading role in space traffic management and advocates for increased funding and safety measures for spacecraft.* |
| AI-92 | *Vice President Kamala Harris speaks at NASA Goddard Space Flight Center in November 2021.* |
| AI-93 | *The page discusses the need for bold policies to support the expansion of human civilization into space, highlighting the role of space-based technologies in various areas. It mentions specific actions and recommendations pursued by the Biden Administration and Congress to strengthen U.S. competitiveness and address challenges in the space industrial base.* |
| AI-94 | *The page discusses the importance of funding, standards, and regulations in the space industry. It highlights the role of the U.S. commercial space industry in international engagement but also raises concerns about foreign ownership and control. The current state of the industry is growing, but inflation and supply chain issues are impacting funding.* |
| AI-95 | *Space startups are experiencing a decrease in funding, but the importance of space-based technologies remains high, particularly in countering Russian aggression in Ukraine. The Biden Administration continues to support space exploration and commercial partnerships, such as the launch of the CAPSTONE spacecraft.* |
| AI-96 | *The page discusses the efforts of the US government to support the space industrial base through financing vehicles and regulatory improvements. It highlights the need for coordination and a coherent investment policy to ensure long-term competitiveness.* |
| AI-97 | *The lack of patient capital in the private sector is hindering the development and commercialization of space assets. The US government historically provided tools to encourage investment in long-term technologies, but the regulatory framework and export controls need streamlining and transparency.* |
| AI-98 | *The page discusses the need for regulatory reforms in the space industry to keep up with technological innovation, the importance of treating space activity as a priority export market, and the potential impact of government investment on the competitiveness of the US space industry.* |
| AI-99 | *The page discusses the concept of a fast follower strategy for adopting commercial technologies within the Department of Defense, particularly in the space industry. It emphasizes the need for a dedicated organizational home for commercial space technology and highlights the benefits of procuring commercial solutions over traditional government acquisition programs.* |
| AI-100 | *The page discusses recommendations for the development of a commercial space roadmap and the elevation of the Office of Space Commerce to facilitate the execution of a national space investment policy. It also suggests optimizing SBIC programs to target capital-intensive space technologies.* |
| AI-101 | *The page discusses the need for increased transparency and communication in the regulatory application process for space companies. It also suggests reviewing and updating the list of restricted exports and coordinating US government investment tools and interventions in the space industry.* |
| AI-102 | *The page discusses various recommendations for the development and improvement of the space industry, including funding challenges, economic models, regulatory processes, infrastructure investment, supply chain development, and international engagement.* |
| AI-103 | *DFARs should be simplified to make it easier for commercial companies to do business with the USG/DoD. Space should be designated as an economic opportunity zone to encourage growth and investment. The planning and* |

AI-105 | *permitting of space infrastructure should be deferred to state and local governments.*

AI-105 | *China's space industry is growing rapidly, with a larger and faster-growing STEM workforce compared to the US. To compete in the second space race, the US needs to prioritize STEM education and attract more talent to support space exploration and defense. China's advantage lies in its larger population size, which allows for greater research and development capabilities.*

AI-106 | *China's workforce in research and development (R&D) is surpassing that of the United States, with a threefold population advantage contributing to the gap. China is projected to have significantly more STEM graduates than the US by 2030, highlighting the need for the US to strengthen its STEM education system to remain globally competitive. The state of the space industrial base relies on a highly educated STEM workforce, which is currently threatened by workforce issues.*

AI-107 | *The page discusses the State of the Space Industrial Base in 2022, highlighting the need for a space-ready workforce and increased investment in STEM education. It also mentions the release of the National Space Council's Space Priorities Framework and the fragmented nature of efforts to attract a skilled workforce.*

AI-108 | *The space industry faces challenges in recruiting and retaining skilled professionals, particularly in mid-level positions. The lack of mentorship opportunities and restrictions on international talent acquisition contribute to the problem. Efforts to diversify the talent pool and address these issues are underway, with some companies pledging to advance diversity and inclusion.*

AI-109 | *A rigorous, six-week, in-resident STEM program with mentors has been found to increase college enrollment, graduation rates, and STEM degree attainment for underrepresented youth. The program provides access to STEM careers, exposure to high-achieving peers and mentors, and immersive science and engineering classes. More programs and mentors like this are needed.*

AI-110 | *The space industry faces challenges in attracting and retaining a skilled workforce due to lack of awareness, education access, hiring requirements, funding gaps, and generational issues. There is a need for more space-focused programs and initiatives to develop talent, as well as immigration reform to retain foreign STEM graduates. Early exposure to STEM is crucial for fostering interest in space careers.*

AI-111 | *The page discusses challenges facing the space industry's workforce, including lack of diversity, retention issues, inequitable access to education, financial gaps, and the need to adjust recruitment strategies for younger generations.*

AI-112 | *The page discusses the state of the space industrial base in 2022, highlighting key inflection points such as the increasing STEM-Graduate gap with China and the need for financial backing and removal of recruitment barriers to grow the space workforce. It also recommends creating a space workforce pathway and emphasizing outreach and support for diverse populations.*

AI-113 | *The page discusses the need for strategic messaging and workforce development in the space industry to attract and retain a diverse workforce. It also emphasizes the importance of aligning STEM education and addressing funding gaps to support the growth of the space workforce.*

AI-114 | *Commercial space companies, in partnership with the government, are driving innovation and shaping the future of space for millions of Americans.*

AI-115 | *NASA's James Webb Space Telescope has provided stunning images and inspired STEM students. The report suggests lower-cost launch options, improved power and propulsion, and in-space servicing for future space exploration.*

| | |
|---|---|
| AI-116 | *The page discusses the potential of building larger and more cost-effective telescopes in space, the importance of unity and aspiration in achieving a better future, and the need for America to lead in space exploration and settlement. It also includes a quote from President Joe Biden emphasizing the possibilities that lie ahead.* |
| AI-117 | *The page lists the participants of a workshop on the state of the space industrial base in 2022.* |
| AI-118 | *The page lists participants and organizations involved in the State of the Space Industrial Base 2022 event.* |
| AI-119 | *A list of individuals and organizations involved in the space industry, including government agencies, consulting firms, and aerospace companies.* |
| AI-120 | *State of the Space Industrial Base 2022 features contributions from various industry experts discussing the current state of the space industry.* |
| AI-122 | *State of the Space Industrial Base 2022 discusses threats, challenges, and actions related to the US space industry. Space Power Competition in 2060 explores challenges and opportunities in the future of space.* |
| AI-123 | *The page discusses key actions and recommendations from the State of the Space Industrial Base 2021 workshop. It highlights the need for a national vision for space development, incorporating the Moon into the Earth's economic sphere, sustaining funding for the Hybrid Space Architecture, and expanding the Artemis Accords beyond NASA. Updates are provided on the progress made in these areas.* |
| AI-124 | *The page discusses the need for international norms in space behavior, increased funding for space science and technology, policy reform, declaring space as a special economic zone, recognizing space as critical infrastructure, and incorporating space into climate action plans. Updates on the progress of these initiatives are provided.* |
| AI-125 | *The page discusses various updates and recommendations for the US space industry, including the need to prioritize space in supply chain planning, integrate JADC2 with the Hybrid Space Architecture, include commercial solutions for in-space logistics infrastructure, increase the percentage of commercial services buys, and expand the use of space commercial services within the Space Force.* |
| AI-126 | *The page discusses the need for acquisition reform, shifting resources from SBIRs to OTAs, investment beyond LEO, and expanding investments in enabling technologies in the space industry. Updates highlight the importance of supporting small businesses and addressing gaps in on-shore manufacturing.* |
| AI-127 | *The page provides the results of a survey conducted during the SSIB'22 Workshop, where participants rated progress on various recommendations for the space industry. The average scores are listed for each recommendation.* |
| AI-128 | *The page discusses recommendations for the space industrial base, including enabling in-space logistics infrastructure, mandating commercial service buys, expanding use of commercial services, promoting acquisition reform, and increasing investment in enabling technologies.* |
| AI-129 | *Attendees of the State of the Space Industrial Base 2022 were disappointed with the progress made in various areas, including expanding investment in enabling technologies and incorporating the Moon into the economic sphere. The most important recommendations to address were recognizing space as critical infrastructure and expanding investment in enabling technologies. These were also considered the easiest to accomplish.* |
| AI-130 | *Blank page.* |

| | *This page provides a list of acronyms and abbreviations related to the space* |
|---|---|
| *AI-131* | *industry, including organizations, technologies, and programs.* |
| *AI-132* | N/A |
| *AI-133* | N/A |
| *AI-134* | N/A |
| *AI-135* | N/A |
| *AI-136* | N/A |

# Notable Passages

AI-1
*From the very beginning, the innovation, the creativity, and the drive of commercial space companies, combined with the resources and the vision of the United States government, has powered America progress in space and our leadership of progress in space.* VICE PRESIDENT KAMALA HARRIS, 2022

AI-6

AI-7 "The future depends on what we do in the present." MAHATMA GANDHI

"China has a grand strategy for economic and military dominance in the space domain that extends for decades as a national priority, complemented with an information campaign that fuels public interest at home and collaborative relationships abroad. China's success seeks to not only shape humanity's future in space, but a new international order that embraces communism and autocracy here on Earth within the 21st century."

AI-8

"In order to save the planet, we must get off-planet. Advancements in off-world power production, manufacturing, and Lunar resource extraction will be foundational to the future multi-trillion dollar space economy. In order to lead, enabling strategy, policy, and law is required. Activities and human presence in space should be driven by an international"

AI-9

"We must accelerate progress on strengthening the space industrial base. The pace of progress towards past recommendations has not been brisk enough to lead. We must catalyze domestic manufacturing and supply chains to reduce costs and lead times while making headway on recommendations from previous years SSIB recommendation."

AI-10

*We are in a race that we might win, but it is certainly conceivable that China may advance in certain areas of space faster than we do.* U.S. SENATOR JERRY MORAN

AI-11

"The space race of today is not like the one we faced with the Soviets. One can actually argue that the space race today has far more on the line. In this new era where the United States is being challenged across every sector, our American space enterprise cannot afford to lose focus or momentum. Therefore, to ensure the United States remains the global space power, we must commit to a set of pillars that are based on principles." - REP. ROBERT ADERHOLT, 2022

AI-12

"The U.S. is now in the dawn of a new era of strategic competition that threatens its peaceful and prosperous vision for the future on Earth and in space. China's recent emergence as a capable competitor in the space domain is concerning, as is its competing vision wherein the U.S. is permanently displaced as the leading economic and military power on Earth and beyond. China has defined both the timespan and scope of the new space race for sustainability, prosperity and the planet. China's national space timeline is 2045 where it intends to be the foremost spacepower: 1) eclipsing the U.S. in its economic activities and services; 2) eclipsing the U.S. in its foreign partner relations and shaping of the space governance system

AI-13

AI-14 "Were China to win this race to industrialize the economic zone of space and secure its governance through leadership and might, it would have vast consequences for the security of the U.S. and its allies. The international order will have changed significantly."

AI-15 "The U.S. will most likely lose space superiority to China within the next decade."-- ATLANTIC COUNCIL REPORT (2022)

AI-16 "Last year's report provided extensive guidelines for what the community would find acceptable and suitably ambitious to ignite national energies. The consensus remains a vision of economic development and settlement -- at least on par with the expansive vision and timeline of our pacing competitor."

AI-17 "China is determined and capable of winning the New Space Race - China is making steady progress toward their goal of surpassing the U.S. as the dominant space power by 2045, with some participants assessing this may happen earlier, unless proactive measures are taken now to sustain our nation's leadership across all instruments of national power."

AI-18 "To explore the vast cosmos, develop the space industry and build China into a space power is our eternal dream." --PRESIDENT XI JINPING

AI-19 "Space is an economic domain that is resource and opportunity rich and provides many manufacturing environments that are too costly or impossible to reproduce on Earth such as ultra-high vacuum, super cold temperatures and microgravity. There has been increasing investment in space companies, particularly startups. Yet, several challenges stand in the way of developing a truly robust space economy."

AI-20 "China continues making steady gains in innovation - Although many U.S. companies choose to keep their intellectual property internal and not ...patents, it is nevertheless concerning that China is making steady gains in patent quality. As assessed by RAND in Figure 8 above, China has made steady gains in Quality-Adjusted Military Patents, surpassing the U.S. since 2018."

AI-21 An American Comeback is Possible - Despite China's recent gains, several long-term trends argue a U.S. comeback is possible. This year, the U.S. economy is forecast to grow at a faster rate than China's for the first time since 1976. China's national security innovation base relies on aging, foreign-trained talent and faces shortfalls in workers with needed skills. China is projected to reach its Lewis Turning Point within the next decade."

AI-22 "In the absence of a clearly articulated North Star vision that focuses efforts, streamlines bureaucratic processes, and energizes the public sector, the strategic end state of the U.S. retaining its leadership role in space is unwittingly compromised."

AI-23 "The Chinese government steals staggering volumes of information and causes deep, job-destroying damage across a wide range of industries - so much so that, as you heard, we're constantly opening new cases to counter their intelligence operations, about every 12 hours or so." - CHRISTOPHER WRAY, Director FBI, 2022

AI-24 "In the absence of Wall Street leadership, other nations are strengthening and diversifying finances of the new space economy."

AI-25 ,"The advantage America enjoys today stems from our space industrial base. We must work together to ensure that it remains a strong, effective, and innovative partner in sustaining American space superiority. The 2020 Defense Space Strategy recognizes that commercial space activities have expanded significantly in both volume and diversity."--GEN JOHN W. RAYMOND, First Chief of Space Operations

| | |
|---|---|
| AI-26 | "In order to save the planet, we must get off-planet. Advancements in off-world power production, manufacturing, and Lunar resource extraction will be foundational to the future multi-trillion dollar space economy. In order to lead, enabling strategy, policy, and law is required. Activities and human presence in space should be driven by an international rules-based order and systems that uphold liberty and prosperity for all humankind." |
| AI-27 | "This tiny Earth is not humanity's prison, is not a closed and dwindling resource, but is in fact only part of a vast system rich in opportunities, a High Frontier which irresistibly beckons and challenges the American genius." |
| AI-28 | n/a |
| AI-29 | "Exploration by a few is not the grandest achievement. Occupation by many is grander." |
| AI-30 | "We are witnessing a second space race. This time the primary components are not status-seeking for a global audience, but lasting economic power. It is the beginning stage of a competition to secure positional, logistical, and industrial advantage. It is a competition to secure the material wealth of space resources, and to convert that wealth into power to shape the international system." |
| AI-31 | "America must be aware that the autocratic space powers are bringing a strong ground-component to the new space race, aggressively seeking space partnership agreements on planet Earth. As of May 2022, the think tank PSSI identified globally 303 Chinese and Russian transactions targeting 83 countries (see Figure 16). Fourteen of those transactions are between China and Russia. Out of the total number of recorded bilateral transactions, China accounted for 147 transactions targeting 71 countries. Russia for 130 transactions targeting 43 countries. Twelve additional international multilateral agreements were sponsored by Russia and/or China." |
| AI-32 | "Time is running out. We have, at most, two years before founders walk away and private capital dries up. And many, many startups will go out of business waiting for DoD to award real production contracts."--KATHERINE BOYLE, GP, Andreessen Horowitz |
| AI-33 | "It's Still Not Enough. While the above are important actions, workshop participants still assessed that U.S. policy was underperforming in supplying the necessary vision and framework, especially when compared with China's Dream, and that the absence of such a North Star Vision constituted the major bottleneck in maintaining enduring U.S. advantage. There is a need to cast a broader, more inspirational and inclusive vision for all Americans. There is consensus that it must focus on economic development and prosperity with the goal of an in-space human-centric economy. The Space Priorities Framework is an important first step, but a stronger connection between space and America's more important challenges, a more inspirational quality, and a set of ambitious goals or timelines would further strengthen the vision. The document |
| AI-34 | "Choosing to remain terrestrially rooted with spiraling populations and industrialization is not a viable alternative. Moreover, China is on track to exceed the U.S. in science and technology investment by 2030 -- the U.S.'s share of global R&D spending declined 2% between 2010 and 2019, while China's share increased 7%. and U.S. federal investment in the science and technology needed to spark innovations crucial to the future of the U.S. economy and national security has fallen to 0.7% of our GDP today compared to 1.9% in 1964." |
| AI-35 | "We need to protect Earth and the way we will protect it is by going out into space." --JEFF BEZOS |

AI-36  "Human civilization is at a new moment of transition across social norms ,economics, governance, and the environment, and is facing the dawn of a new era of inter-planetary human migration (to Mars).

AI-37  "The commercial space industry is also a powerful engine for economic growth. Our nation's space economy employs over 354,000 people and generates $200 billion a year." - VICE PRESIDENT KAMALA HARRIS, 2022

AI-38  "Today, while the public may not fully realize it, the space ecosystem touches many aspects of daily life, abounds with commercial investment and increased commercial use cases, and plays a key role in advancing global sustainability and security priorities. It holds even greater potential for the future."-- McKINSEY & COMPANY

AI-39  "The fatalism of the limits-to-growth alternative is reasonable only if one ignores all the resources beyond our atmosphere, resources thousands of times greater than we could ever obtain from our beleaguered Earth." - DR. GERARD K. O'NEILL 1978

AI-40  "To explore the vast cosmos, develop the space industry and build China into a space power is our eternal dream," stated President Xi Jinping. The space industry is a critical element of the overall national strategy, and China upholds the principle of exploration and utilization of outer space for peaceful purposes. Since 2016, China's space industry has made rapid and innovative progress, manifested by a steady improvement in space infrastructure, the completion and operation of the BeiDou Navigation Satellite System, the completion of the high-resolution earth observation system, steady improvement of the service ability of satellite communications and broadcasting, the conclusion of the last step of the three-step lunar exploration program (,"orbit, land, and return"), the first stages in building the space station, and a smooth

AI-41  "Why the moon? Because the goal is Mars. What we can do on the moon is learn how to exist and survive in that hostile environment and only be three or four days away from Earth before we venture out and are months and months from Earth. That's the whole purpose: we go back to the moon, we learn how to live there, we create habitats."--SENATOR BILL NELSON, NASA Administrator

AI-43  "The U.S. has a healthy lead with the evolution of reusable launch vehicles and alternatives to traditional launch such as horizontal (airborne) and kinetic (spin) first-stage equivalents. That said, it is ill-advised to slow or stop the innovation that improves and enhances these capabilities. The space economy is driven by the downstream revenues, but the upstream segment, including launch services, is the backbone of the sector. No nation can maintain a dominant position in space without a superior capability for rapid movement and sustainment. The marketplace is ever changing and demands continuous innovation. This New Space Race will be more contentious and more complex than the first, involving a broad spectrum of powerful commercial interests, and with much more at risk for America."

AI-44  "Innovation is Reducing the Cost per Kilogram to Orbit - The scope and diversity of launch vehicles (and alternatives to traditional methods of accessing space) hold great potential to disrupt the launch services market. Multiple launch providers are developing and testing reusable launch vehicles while others leverage horizontal (airborne) or kinetic launch as alternatives to first stages. Collectively, this innovation is driving the cost per kilogram lower by a factor of five or more. The resulting impact is closing business cases for new on-orbit services and growing the new space economy."

| | |
|---|---|
| AI-45 | "Growing Traffic Congestion in the Air and at Sea - Launch and landing activities will increasingly affect other existing transportation networks. On the Eastern and Western ranges, the shutting down of the air and water space for every launch affects airline and mariner's routes, as well as, general aviation and recreational boating, whether the launch attempt is successful or not. These new activities will eventually impact other terrestrial transportation of roads & rail as the spaceport system evolves across the country. Real conflicts are arising with every launch activity at the Cape, and they will inevitably occur elsewhere." |
| AI-46 | "What's Driving Range of the Future - The current capacity of U.S. space launch complexes is insufficient to meet the growing demand by U.S. commercial, civil and national security launch providers. If nothing changes, the U.S. will reach a choke point where the growth and responsiveness of launch operations will be artificially curtailed creating greater economic opportunity for strategic competitors (i.e., China and Russia). Launch, landing, and payload processing demands are up, stressing the availability of existing assets and facilities. Both Eastern and Western ranges, in particular, are seeing those demands increase rapidly, highlighting current infrastructure restrictions due to aging facilities and constrained readiness under existing operational procedures and policy. According to the USSF, the average age of range items with a potential mission |
| AI-47 | "There are several concepts being evaluated by different parties to address the same fundamental challenge. One option under consideration by the USSF PEO for Space Launch is to address funding shortfalls for maintenance and to support upgrades to newer capability. This would require new authorities from Congress to allow USSF to accept industry partner reimbursements to jointly fund infrastructure improvements (a Public Enterprise Revolving Fund)." |
| AI-48 | "There is no greater challenge for the Eastern and Western ranges, and their government partners with landlord or regulatory responsibilities, than to transition away from the traditional mission assurance culture. That mentality had its purpose and was instrumental in aiding U.S. leadership in the early days of space transportation. But as history teaches us again and again, disruptive innovation is how leadership is sustained. The commercial sector is now driving U.S. leadership once again through rapid iteration and agility; where failure and learning from it are foundational elements of their culture for success." |
| AI-49 | " The first fully reusable launch vehicle ...successfully from a U.S. spaceport demonstrating the viability and compatibility of performing developmental test and evaluation operations at commercial speed, and from someplace other than a high-density spaceport." |
| AI-50 | "U.S. leadership in space will diminish under the burden of regulation and a bureaucracy that is ill-suited for today's strategic competition. Innovation and agility in policy making and regulating is required to meet the evolving demands of the 21st century. Appropriate waivers and exclusions should be exercised or considered, when necessary, in the best interest of U.S. economic growth and national security to remain competitive in this new frontier. Public safety, however, must not be compromised." |
| AI-51 | "As we bring more and more commercial providers on board, we're going to have more and more launches to get to that ultimate goal of two a day. It's all about converting our thinking and processes into an airport or services model in order to get there." - BRIG GEN STEPHEN G. PURDY, Space Launch Delta 45 Commander |

| | |
|---|---|
| AI-52 | "Range of the Future postures America to sustain its leadership in the new space economy and ultimately to enabling humanity to become a multi-planet species." |
| AI-53 | ,"US leadership was foundational in establishing the global air marketplace. As a result, every international air traffic control tower across the planet uses, of necessity, a common language; English. In space there is no such common language for humanity to conduct business, yet. Space is hard enough, without having to do it in Mandarin." - DALE KETCHAM, Space Florida, 2022 |
| AI-54 | n/a |
| AI-55 | ,"The committee is pleased that most national security space organizations have publicly embraced the Hybrid Space Architecture concept, notably the Space Force, National Reconnaissance Office, National Geospatial-Intelligence Agency, and the Space Development Agency. However, the committee further notes that funding for the Hybrid Space Architecture has historically lagged in budget submissions" |
| AI-56 | "SpaceX Starlink launched 989 new satellites in 2021 across 19 launches, increasing their active subscriber base by 400%. Starlink further announced partnerships with Microsoft Azure and Google Cloud Platform, and proved itself rapidly deployable and surprisingly resilient to both jamming and cyber attacks throughout the Ukraine conflict." |
| AI-57 | "The Ukrainian conflict highlights the resilience and vulnerability of commercial communications - Early in the conflict, most Viasat ground terminals were rendered inoperable by Russian activities, but within days, over 15,000 SpaceX Starlink terminals were shipped to Ukraine and rapidly deployed reestablishing effective communications within the country despite Russian jamming and hacking attempts. Further, wearable 5G pucks with local mesh networking and short databurst satellite links drastically enhanced tactical situational awareness to field operators. It is evident to U.S. adversaries that commercial communications assets (in space and on the ground) are strategically and tactically important, and thus going forward, we can expect measures will be employed by adversaries to disrupt these assets." |
| AI-58 | "Dual-use technologies - how will the government, and especially the military, adapt to trusting commercially owned and commercially operated networks to transport sensitive data at varying levels of classification, potentially including weapons targeting information? Is industry willing to carry sensitive government data, potentially including weapons targeting information?" |
| AI-59 | "The Hybrid Space Architecture will dramatically improve deterrence and resilience in space while providing substantial new information advantage for science, commerce, and security." |
| AI-60 | "Long term leadership and funding - As identified in the chapter featured quotation, there have been enthusiastic endorsements from leadership but lagging funding thus far. It was acknowledged that SWAC has been assigned some funding through a Program Decision Memorandum (PDM)." |
| AI-61 | "DTN is a set of standard protocols that make use of data streams and multi-path communications to successfully deliver information through a number of network nodes. DTN addresses this by using a Bundle Protocol (BP) whereby each node can "store and forward" streams without disruption if errors or disconnections occur. In this scenario, data streams literally "hop" reliably from one node to the next. DTN is beneficial not just for human missions like Artemis, but for robotic missions as well. It's the first step in establishing the Outernet, a commercial internet in space, and creating new economic opportunities for all." |

| | |
|---|---|
| AI-62 | "The Committee supports efforts to leverage commercial space networks to create an "outernet" for future military communications and believes the Space Force should undertake activities to promote interoperability standards and use of commercial ground and cloud architectures to increase the integration of commercial space networks."--HOUSE ARMED SERVICES COMMITTEE, 2022 |
| AI-64 | n/a |
| AI-65 | "The United States recognizes the importance of establishing an ecosystem that promotes an affordable and sustainable (space) transportation network and logistics capability." --NATIONAL ISAM STRATEGY, 2022 |
| AI-66 | "The U.S. government, industry, and academia are all pushing on the ISAM front and there have been recent significant achievements." |
| AI-67 | "Foreign Powers Continue to Make Bold Moves - Other countries are making strides towards developing ISAM capabilities, including ambitious programs such as the European Union's Prototype for an Ultra Large Structure Assembly Robot (PULSAR) and the MOdular Spacecraft Assembly Reconfiguration (MOSAR) Horizon 2020 projects. China's Shijian-21 towed a dead satellite to another orbit. Astroscale, a Japanese company, End-of-Life Services by Astroscale-demonstration (ELSA-d) successfully demonstrated its capability to repeatedly magnetically capture a client satellite. The U.S. cannot assume that it will be the unchallenged leader in the future logistics chain." |
| AI-68 | ,"Economic growth requires innovation. Trouble is, Washington is practically designed to resist it. Built into the DNA of the most important agencies created to protect innovation, is an almost irresistible urge to protect the most powerful instead."--LAWRENCE LESSIG, Harvard Law School, 2008267 |
| AI-69 | "Imagine flying a commercial airliner to Europe and then throwing it away at the end of the journey. This is a simple but effective analogy to describe the challenges in the early days of space operations due to the tyranny of the rocket equation which mathematically tells us that generally no more than 4% of an expendable rocket's total mass is usable for its intended payload." |
| AI-70 | "Strategic competitor achieves logistics chain dominance first. If the U.S. does not lead in ISAM and other countries keep advancing, the U.S. might be beholden to the ways of practice, standards, and interfaces adopted by the larger space community first, stifling innovation and growth of the U.S. industrial base." |
| AI-71 | "The U.S. leads the modernization of space laws and treaties, most of which did not contemplate commercial activity in space. The U.S. can either seek to reform existing frameworks or forge new multilateral agreements for rules of engagement, such as the Artemis Accords." |
| AI-72 | n/a |
| AI-73 | ,"We keep starting and stopping. We may have spent as much as $20 billion on developing (but never buying except once) space ... systems."--BHAVYA LAL, NASA |
| AI-74 | "Next generation power technologies are needed to enable spacecraft to run even more demanding payloads (e.g., directed energy) at higher duty factor to maximize both military and commercial utility." |
| AI-75 | "The DARPA NTP program expects to launch a demonstration vehicle by 2026, while NASA's higher performance NTP concept may be demonstrated towards the end of the decade, benefiting from the regulatory, safety and licensing pathway forged by the DARPA program. These two programs are expected to make major headway on remaining technical challenges associated with NTP (e.g., high temp materials, long duration cryogenic storage) in the next several |

| | |
|---|---|
| | *years, possibly opening up new commercial opportunities in nuclear thermal propulsion."* |
| AI-76 | *"Power Beaming and Space Based Solar Power - Commercial Photovoltaics (PV) are seeing interest in very high power arrays, possibly exceeding Megawatt (MW)-scale power output. Deployment and/or assembly of MW-scale solar arrays and associated technologies (e.g. high-voltage power processing) will enable space-based power beaming, but large uncertainty exists around the business case and demand signal for space-to-Earth beamed power. Technical progress has been made on converting solar to RF on the meter-scale and transmitting this power in the lab (AFRL SPIDR), although the U.S. is still lagging behind Chinese eÔøΩorts towards large scale SBSP demonstrations, with the completion of a 75-m test tower at Xidian* |
| AI-77 | *"We forecast that the highest growth in the market will be from new space applications and industries, which will be unlocked as the industry becomes more affordable, accessibility becomes more widespread, and technology improves. This includes areas such as space-based solar power, Moon/asteroid mining, space logistics/cargo, space tourism, inter-city rocket travel, and microgravity R&D and construction" -- CITIGROUP* |
| AI-78 | *"If you stop and think about it, the form of propulsion used today hasn't changed in over a thousand years... since the invention of fireworks by the Chinese. Basically you burn (oxidize) a material in a tube, hot gasses come out one end and the vehicle Ô¨Çies in the opposite direction. Sure our rockets have gotten bigger and more efficient, but the basic design remains unchanged."--DR. PETER DIAMANDIS, 2010* |
| AI-79 | *"The quest for safe fusion and other reactors will not only improve life in space, but are essential to enable humanity's evolution from fossil fuel energy sources harmful to Earth's unique biosphere. In space, power is life; and reliable continuous sources of power will prove pivotal to building and sustaining an off-planet economy and future human settlements."* |
| AI-80 | *"Starship reaches orbit demonstrating 100-ton to LEO - would be a stepwise change in launch cost and access to space. The impacts to commercial business plans and how this will enable new concepts of operation in Earth orbit are still being understood. A successful Starship orbital demo may also inform feasibility of high-mass space based solar power (SBSP) beaming projects and MW-scale nuclear fission reactor space propulsion and power concepts."* |
| AI-81 | *"First, I believe that this nation should commit itself to achieving the goal, before this decade is out, of landing a man on the moon and returning him safely to the Earth... Secondly, an additional 23 million dollars, together with 7 million dollars already available, will accelerate development of the Rover nuclear rocket. This gives promise of someday providing a means for even more exciting and ambitious exploration of space, perhaps beyond the moon, perhaps to the very end of the solar system itself."- PRESIDENT JOHN F. KENNEDY, 1962* |
| AI-84 | *"It has brought glaring attention to the capability of commercial remote sensing companies to provide a timely record of war events and casualties, and the economic consequences flowing from the same. Commercial remote sensing data therefore strengthens the 'Information' element of U.S. National Power by further enabling detection and subsequent prediction of military maneuvers, and then distributing that information to people across the globe who may be contending with misinformation and disinformation."* |
| AI-85 | *"A Mixed Bag of Positive Outcomes with Growing Concerns - The last 12 months has demonstrated both progress and stagnation. NRO's award of multiple contracts* |

to remote sensing companies is a positive step toward providing meaningful adoption of these services, while some looming concerns over the level of commitment and consistency of that adoption, licensing and distribution restrictions, and supply chain persist. This 'mixed bag' sets the stage for a number of the inflection points detailed further below, but also presents opportunities to set the U.S. and the space industrial base on a stronger course for success and increasing prosperity."

AI-86 "Commercial imagery has proven invaluable in the Ukraine conflict, both to the U.S. government's efforts to counter the Russian information war, as well as to the Ukrainian military."

AI-87 "Without addressing the space domain awareness challenge identified above, and without allocating a budget in line with the scale of this need, the U.S. cannot effectively sustain its leadership in space traffic management across all altitudes and regions."

AI-88 "When Russia invaded Ukraine in 2014, the unexpected events led one BBC reporter to observe, 'the annexation of Crimea was the smoothest invasion of modern times. It was over before the outside world realized it had even started.' Eight years later, commercial space technology has profoundly changed the character of conflict. Months before this February's invasion of Ukraine, commercial remote sensing satellites provided early indications and warning of Vladmir Putin's sinister intentions. Armored columns literally had no place to hide regardless of day, night or inclement weather conditions."

AI-89 "China establishes dominance in the global remote sensing market - by number of sensors, and is quickly adding customers through simple purchasing mechanisms. The U.S. retains the crown for fidelity and data quality, but at higher prices and with USG/ITAR-imposed restrictions intact."

AI-93 "While SpaceX and others have been taking bold steps toward paving a technical path toward space settlement, there is an equivalent need to take bold steps toward enacting the necessary policies that will enable the expansion of human civilization into the solar system and beyond." - DR. SIMON ,"PETE,"WORDEN, (Brig. General, USAF, Ret.)

AI-95 "The technologies and capabilities being developed in the space industrial base, however, remain more important than ever. Perhaps most notably, the confli in Ukraine underscores the importance of dual-use space-based technologies, such as commercial satellite imagery, which is enabling real-time insight into Russian military activities. The reliable and multiple sources of truth are effectively countering attempts by the Kremlin to promote false narratives and providing significant ballast to Ukraine, the NATO alliance and the global community in turning back this unwanted aggression."

AI-96 "In addition to private sources of financing, the USG is responding to the call from industry to pursue creative financing vehicles to seed the space industrial base. In particular, the U.S. Export Import Bank launched the Make More In America Initiative which, when combined with ExIm's China & Transformational Exports Program (CTEP), is designed to unlock loans and loan guarantees to 'deep tech' industries such as space, enabling U.S. companies access to low-cost capital in order to scale manufacturing in the U.S."

AI-97 "Lack of Patient Capital - Stakeholders from across the U.S. commercial and national security space ecosystem identify the lack of patient capital in the private sector as a significant bottleneck to developing, maturing and commercializing dual-use space assets."

| | |
|---|---|
| AI-98 | "U.S. pursues regulatory reforms that align with the pace of technological innovation and diffusion to allow more coordinated and streamlined decision making processes domestically and internationally related to export controls and investment screening." |
| AI-99 | "I keep this chart handy as a reminder why we need cislunar space domain awareness today, and why the USSF needs to "keep up" with the technology, innovation, standards, policy, norms, etc. that this human expansion to the Moon and beyond represents." - DR. JOEL MOZER, USSF |
| AI-100 | "Develop a Commercial Space Roadmap that Executes a National Space Investment Policy - Create a central office or body that would have a 'policy planning' type mandate to develop the commercial space roadmap. This roadmap which would, among other things: identify the technologies requiring funding; identify the resources required to support future industries; Identify the offices across government that would be required to take action in a coordinated manner to disburse the resources and enable the industries. This office should be at the Office of Space Commerce at the Department of Commerce; this office would need to be elevated out of NOAA and become a direct report to the Secretary, be provided with greater resources, and a mandate to direct work across the government." |
| AI-101 | Increase Transparency and Communication in Regulatory Application Process - Many companies find the U.S. regulatory system cumbersome, slow, and opaque. Some USG agencies have implemented metrics for approvals and provide tracking status via a portal for regulatory and license review and approvals. However, if the approval deviates from the standard process, which they often do, the portal and existing metrics are no longer used. This leads to frustration in the industry and lack of confidence in the process. At this point, transparency is lost between the government and industry. The lack of transparency can further delay approvals, as simple questions or clarifications to the requestors may be unresolved for days. (OPRs: DoC/BIS, DoD/DTSA |
| AI-102 | "Establish Single Licensing Agency or Single Face to the Customer - The U.S. regulatory process is exceedingly slow and opaque, stifling innovation domestically and between the US and international partners. In general, there is currently not a single point of approval or denial, and it is often difficult to achieve consensus between the necessary regulatory agencies (e.g., FAA, FCC, Dept of State, Dept of Commerce). Any agency can delay or deny approvals. This process lacks transparency and takes a "red light" approach to approvals. Establishing a single licensing agency will enable a 'single point of contact' for industry, concentrate government regulatory expertise, and accelerate the process and transparency." |
| AI-103 | USG should designate space as an economic opportunity zone to encourage growth and investment - The U.S. can designate Space as an economic opportunity zone in order to improve the economic prospects for critical space-technologies or infrastructure. Historically, U.S. economic zones oÔøΩered a range of beneÔøΩts, including favorable tax treatment for investors and companies, improved trade opportunities, or the ability to leverage federal infrastructure. In order to advance U.S. space competitiveness, policy makers need to consider such a designation in order to incent stakeholders across the space community to actively invest in space. (OPRs: Treasury, OMB, Commerce, Congress) |
| AI-104 | n/a |
| AI-105 | "China has made space a key element of their national strategy and their progress is evident in the increasing strength of their science, technology, engineering and |

| | |
|---|---|
| AI-106 | math (STEM), or STEAM with the arts included, workforce --which outpaces all nations in growth and size." |
| | "The United States needs to reinvigorate its STEM education system if it is to compete successfully in the 21st century. STEM proÔ¨Åciency has been declining in America since the 1980s, threatening the nation's continued technological leadership." --Center for Strategic and International Studies (CSIS), 2022 |
| AI-107 | "Across the U.S., more than half of high schools do not oÔ¨Äer calculus, four out of ten do not oÔ¨Äer physics, more than one in four do not oÔ¨Äer chemistry, and more than one in Ô¨Åve do not oÔ¨Äer Algebra II (a gateway class for STEM success in college)."--Center for Strategic and International Studies (CSIS), 2022 |
| AI-108 | "The U.S. is at risk of losing $454 billion of manufacturing GDP in 2028 alone due to a skills shortage."--DELOITTE and the MANUFACTURING INSTITUTE, 2018361 |
| AI-109 | "Students offered seats in the STEM summer programs are more likely to enroll in, persist through, and graduate from college, with gains in institutional quality coming from both the host institution and other elite universities. The programs also increase the likelihood that students graduate with a degree in a STEM field, with the most intensive program increasing four-year graduation with a STEM degree attainment by 33 percent." |
| AI-110 | "Many, if not most, of the foreign nationals earning advanced STEM degrees from U.S. universities would prefer to stay and work in the country. But America's immigration system is turning away these workers in record numbers ‚Äì and at the worst possible time."--POLITICO, 2022 |
| AI-111 | "Although Blacks or African Americans, Hispanics or Latinos, and American Indians or Alaska Natives represent 30% of the employed U.S. population, they are 23% of the STEM workforce due to underrepresentation of these groups among STEM workers with a bachelor's degree or higher."--NATIONAL SCIENCE FOUNDATION, 2021 |
| AI-112 | "One thing business leaders and educators readily agree on is that if we are to have sustained growth in the space industry, we must have an uninterrupted pipeline of talent. The jobs available in the global space ecosystem are becoming more varied and increasingly technical in nature and are destined to help create new products and services both in space and on Earth. If we are to realize that growing potential, we must have the talent pool to get us there."--THOMAS "TOM" ZELIBOR (Rear Admiral, USN, ret), CEO, SPACE FOUNDATION |
| AI-113 | "Create strategic messaging at a national level that can be used by all regions and sectors of the industry to showcase the full range of job opportunities in space, and past successes from minority individuals in space careers, to attract and retain a larger, more diverse workforce." |
| AI-114 | "Today, at laboratories, on launchpads, and in orbit ‚Äì often, in partnership with our government'a commercial space companies are making real the opportunity of space for millions of Americans. Their work is accelerating innovation in the space sector and shaping our nation's future in space."--VICE PRESIDENT KAMALA HARRIS, 2022 |
| AI-116 | "What America needs now is a mission, a purpose that strikes at the selfless soul of both young and old. We must lead. We must prevail. We must trailblaze a path to the Moon, Mars and beyond that creates hope and opportunity for all humankind. Much as America has been the beacon of hope for millions of emigrants throughout our past, the solar system provides new worlds and new opportunities for potentially billions of people in the future. America needs to |

AI-124 "shape its space future today in order to assure future generations the opportunity to lead from the front in shaping the next great adventure for humanity in the future. A North Star Vision of economic development and human settlement is the first step toward achieving that goal."
"Space Policy Directive 1 directed NASA to: 'Lead an innovative and sustainable program of exploration with commercial and international partners to enable human expansion across the solar system and to bring back to Earth new knowledge and opportunities. Beginning with missions beyond low-Earth orbit, the United States will lead the return of humans to the Moon for long-term exploration and utilization, followed by human missions to Mars and other destinations.'"

AI-125 n/a

AI-126 "Balanced Growth Requires Investment Beyond LEO - The innovations in venture capital and SPACs have put the Low Earth Orbit (LEO) economy on solid ground with a tremendous diversity of space access options and space information services. However, we do not yet have a MEO, HEO, GEO, Cislunar or Lunar economy that is new space-oriented."

# STATE OF THE SPACE INDUSTRIAL BASE 2022
Winning the New Space Race for Sustainability, Prosperity and the Planet

Summary Report by:

## J. OLSON,[1] S. BUTOW,[2] E. FELT,[3] & T. COOLEY[4]

[1]United States Space Force, [2]Defense Innovation Unit, [3]Department of the Air Force and [4]Air Force Research Laboratory

Edited By:
**PETER GARRETSON**

August 2022

DISTRIBUTION STATEMENT A. Approved for public release: distribution unlimited.

# DISCLAIMER

The views expressed in this report reflect those of the workshop attendees and do not necessarily reflect the official policy or position of the U.S. Government, the National Aeronautics and Space Administration (NASA), the Department of Defense (DoD), the U.S. Air Force (USAF), or the U.S. Space Force (USSF). Use of NASA photos in this report does not state or imply the endorsement by NASA, or by any NASA employee, or the DoD, or by any DoD employee, of a commercial product, service, or activity.

**Cover:** Illustration depicting Starship at stage separation from the Super Heavy Booster while on ascent to low Earth orbit. (Source: SpaceX)

## ABOUT THE AUTHORS

**Major General John M. Olson, Ph.D, USAF**
Mobilization Assistant to the Chief of Space Operations, USSF

**Colonel Eric Felt, Ph.D, USSF**
Director of Space Architecture, Science and Technology
Office of the Assistant Secretary of the Air Force For Space Acquisition and Integration (SAF/SQ)

**Steven J. Butow**
Space Portfolio Director, Defense Innovation Unit
OUSD R&E

**Dr. Thomas Cooley**
Chief Scientist of the Air Force Research Laboratory (AFRL) Space Vehicles Directorate

## ACKNOWLEDGEMENTS FROM THE EDITOR
Peter Garretson

The authors wish to express their deep gratitude and appreciation to Space Florida and NewSpace New Mexico for hosting the State of the Space Industrial Base 2022 Workshop in Cape Canaveral, FL, and Albuquerque, NM; and to all the attendees, whether live or virtual, who spent the time and resources to share their observations and insights to each of the six working groups. The workshop and this report would not have been possible without the dedicated efforts of the working group chairs and co-chairs: Russ Teehan, Chris Paul, Rogan Shimmin, Karl Stolleis, Samantha Glassner, Pav Singh, Katherine Koleski, Barry Kirkendall, James Winter, Ryan Weed, Dave Barnaby, GP Sandhoo, Scott Erwin, Casey DeRaad, Dale Ketcham and Helen Park. Nor without the outstanding contributions of our guest speakers and panelists: Bill Nelson, Bhavya Lal, Mike Brown, Bruce Cahan, Namrata Goswami, Robbie Schingler, Brian Weeden, Mark Jelonek, Rick Tumlinson, Chris Paul, Steve Nixon, Jason Aspiotis, Juli Lawless, John Wagner, Steve Wood, Peter Wegner, Amy Hopkins, Brian Flewelling, John Moberly, Shiloh Dockstader, Lee Steinke, Christos Chrisodoulou, Tom Caudill, Maria Tanner, Megan Crawford, Jared Rieckewald, Cameo Lance, Jim Keravala, Brian Weeden, Mark Jelonek, Lisa Rich, Meagan Crawford and Nicholas Eftimiades. The virtual workshop would not have been possible without the incredible support provided by Scott Maethner, Arial DeHerrera, Erika Hecht, Andy Germain, Jamie Holm, Emily Maethner, Andrew MacKenzie, Joe Pomo, Nicole Sena, Carol Welsch, Zachariah Sena, Garrett Rose, Rex Ridenoure, Jason Wallace, Lauren Rogers, Austin Baker, Nathan Gapp, Dennis Poulos, Debbie Willhart, Ellen Cody, Elizabeth Loving, Kelly Dollarhide and Klay Bendle. We also wish to thank David Martin, Ben Felter, Johanna Spangenberg Jones and Ric Mommer for their finishing touches.

## ABOUT THE KEY GOVERNMENT CONTRIBUTORS

### U.S. Space Force | spaceforce.mil

The U.S. Space Force (USSF) is a military service that organizes, trains, and equips space forces in order to protect U.S. and allied interests in space and to provide space capabilities to the joint force. USSF responsibilities include developing military space professionals, acquiring military space systems, maturing the military doctrine for space power, and organizing space forces to present to our Combatant Commands.

### Air Force Research Laboratory | afresearchlab.com/technology/space-vehicles/

The Air Force Research Laboratory's mission is leading the discovery, development, and integration of warfighting technologies for our air, space and cyberspace forces. With its headquarters at Kirtland Air Force Base, N.M., the Space Vehicles Directorate serves as the Air Force's "Center of Excellence" for space research and development. The Directorate develops and transitions space technologies for more effective, more affordable warfighter missions.

### Defense Innovation Unit | diu.mil

The Defense Innovation Unit's (DIU) mission is to accelerate commercial innovation for national security. It does so by increasing the adoption of commercial technology throughout the military and growing the national security innovation base. DIU's Space Portfolio facilitates the Department of Defense's ability to access and leverage the growing commercial investment in new space to address existing capability gaps, improve decision making, enable a shared common operating picture with allies, and help preserve the United States' superiority in space.

**DISTRIBUTION STATEMENT A**. Approved for public release: distribution unlimited.

*This page was intentionally left blank.*

# TABLE OF CONTENTS

| | |
|---|---:|
| EXECUTIVE SUMMARY | 1 |
| INTRODUCTION | 5 |
| THE NORTH STAR VISION: ECONOMIC DEVELOPMENT & HUMAN SETTLEMENT OF SPACE | 23 |
| LAUNCH SERVICES | 37 |
| HYBRID SPACE COMMUNICATIONS | 49 |
| IN-SPACE TRANSPORTATION & LOGISTICS | 59 |
| NEXT GENERATION POWER & PROPULSION | 67 |
| REMOTE SENSING & TRAFFIC MANAGEMENT | 77 |
| POLICY & FINANCE | 87 |
| WORKFORCE & STEM EDUCATION | 99 |
| EPILOGUE | 109 |
| | |
| APPENDIX A - WORKSHOP PARTICIPANTS | A-1 |
| APPENDIX B - PREVIOUS REPORTS | B-1 |
| APPENDIX C - KEY ACTIONS & RECOMMENDATIONS FROM SSIB'21 REVISITED | C-1 |
| APPENDIX D - SSIB'22 PARTICIPANTS SURVEY AND RESULTS | D-1 |
| APPENDIX E - ACRONYMS & ABBREVIATIONS | E-1 |

> "From the very beginning, the innovation, the creativity, and the drive of commercial space companies, combined with the resources and the vision of the United States government, has powered America's progress in space and our leadership of progress in space."
>
> – VICE PRESIDENT KAMALA HARRIS, 2022[1]

Illustration: NASA has selected Axiom Space and Collins Aerospace to advance spacewalking capabilities in low-Earth orbit and at the Moon, by buying services that provide astronauts with next generation spacesuit and spacewalk systems. (Credit: NASA)

> "...the United States of America has no intention of finishing second in space. This effort is expensive — but it pays its own way, for freedom and for America."
>
> – PRESIDENT JOHN F. KENNEDY, 1963[2]

---

[1] White House (2022). Remarks by Vice President Harris On Supporting the Commercial Space Sector. The White House.
[2] Remarks prepared for delivery at the Trade Mart in Dallas, TX, November 22nd, 1963 [Undelivered] as JFK was assassinated while en route to the event.

# EXECUTIVE SUMMARY

Figure 1. Global space power ambitions challenge liberty and prosperity, and these nations are prepared to act decisively in order to achieve their aims. (Credit: Licensed from Getty Images)

*"The future depends on what we do in the present."*
– MAHATMA GANDHI

## GENERAL OBSERVATIONS

Major findings from the workshop include:

- **Strategic competition in space remains a paramount concern** - China continues to compete toward a strategic goal of displacing the U.S. as the dominant global space power economically, diplomatically and militarily by 2045, if not earlier. Proactive measures are required to sustain our nation's space leadership across all instruments of national power despite China's attempt to accelerate closing its technology gap with the U.S.

- **Strategic vision required** - An enduring North Star vision for America in space is an essential guidepost to remain competitive with a rapidly advancing China. Participants recommend 'Economic Development and Human Settlement' as that vision to retain U.S. economic leadership, motivate the American people, and protect U.S. national interests.

- **Space ecosystem at risk** - The agile engineering ecosystem that has become the hallmark of the modern space era is at risk due to some U.S. policy and procurement practices within the bureaucracy that are not aligned with, or work counter to, national space strategy. The national enterprise must adopt and further implement the U.S. National Space Strategy and modify practices to ensure U.S. competitiveness.

- **In order to protect the planet, we must get off-planet** - Advancements in off-world power production, manufacturing and Lunar resource extraction will be foundational to the trillion-dollar space economy. Enabling strategy, policy, and laws will be required to lead in an enlightened and effective effort.

- **Commercial space technology has forever changed the nature of conflict** - Space remote sensing, advanced analytics and broadband communications are tactical solutions that have delivered strategic impact as evidenced by their contribution to the defense of Ukraine.

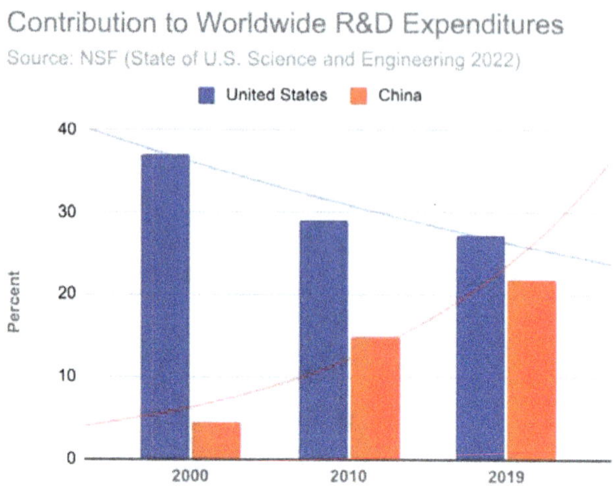

- **Sustaining U.S. leadership in global research and development requires action** - China has rapidly advanced to challenge U.S. leadership in global research and development (R&D) expenditures, and is on track to spend more than $200 billion a year more than the U.S. in R&D by 2030.[4]

Figure 2. U.S. commanding lead in Global research and development has diminished. (Credit: NSF)[3]

**Central message from inputs** - China has a grand strategy for economic and military dominance in the space domain that extends for decades as a national priority, complemented with an information campaign that fuels public interest at home and collaborative relationships abroad. China's success seeks to not only shape humanity's future in space, but a new international order that embraces communism and autocracy here on Earth within the 21st century.

**Major themes from participants** - The U.S. can retain its leadership position in space by accelerating the adoption of commercially-sourced solutions for both civil and national security space applications while reducing the bureaucratic 'roadblocks' that stifle innovation and slow the pace of technological advancement. Countering China's hegemonic ambitions requires innovation and policy reform that is aligned with national priorities of this century rather than the previous one.

**Major opportunities** - Significant private investment continues to be made in the new space economy at record-breaking levels. Continued growth and diversification of this investment requires a strong demand signal from both civil and national security space.

**Major concerns requiring urgent action** - However, while the United States space industrial base remains on an upward trajectory, participants expressed concerns that the upward trajectory of the People's Republic of China (People's Republic of China (PRC)) is even steeper, with a significant rate of overtake, requiring urgent action. The fundamental tonic is to mobilize still greater energies with an enlarged vision and broader set of policy as our nation did in 1962. Specifically, the U.S. lacks a clear and cohesive long term vision, a grand strategy for space that sustains economic, technological, environmental, social and military (defense) leadership for the next half century and beyond. A North

---

[3] NSF (2022). The State of U.S. Science and Engineering 2022. NSF
[4] NSB (2020). Nation Science Board Vision 2030. NSB.

Star vision for economic development and human settlement in space should be bi-partisan, multi-generational, and inspirational to all who embrace America's values.

**A need to act and accelerate progress on previous recommendations** - SSIB'22 workshop attendees were slightly negative (bearish) in their assessment of U.S. progress made over the past year (See [Appendix D](Appendix D)). Efforts and actions must be accelerated to lead.

## KEY ACTIONS & RECOMMENDATIONS

These recommendations reflect the numerous inputs from workshop participants and their best assessment regarding which agency(cies) are in the best position to forward the necessary change.

1. **Establish an enduring U.S. North Star Vision for Space** as an essential guidepost to sustain the United States' competitive advantage against a rapidly advancing China that is now collaborating with Russia. This vision must be as clear and ambitious in scale and timeline as the PRC and more inclusive of international collaboration across the spectrum of commercial, civil and national security space activities. For the first time, participants expressed concerns that China appears to be on track to surpass the U.S. as the dominant space power by 2045 or potentially earlier unless proactive measures are taken now to sustain our nation's leadership. (OPRs: National Security Council (NSC), National Space Council (NSpC), Office of Science and Technology Policy (OSTP))

2. **Enable National Security Priority Processing of Licenses/Environmental Clearances for Critical Space Systems.** The agile engineering ecosystem that has become the hallmark of the new space era is at risk due to U.S. policy and procurement practices put in place by Robert McNamara in 1962 when the government was the source of innovation leadership. Today, we must use Defense Production Act (DPA) Title III and other authorities such as the Defense Priorities and Allocation System (DPAS) to reduce bureaucratic delay on critical needs for national security. (OPRs: NSC, NSpC, OSTP, Office of the Secretary of Defense (OSD))

3. **Elevate the Office of Space Commerce (OSC) to report directly to the Secretary of Commerce.** In a whole-of-nation strategy where commerce and economic growth are central to strategic competition, the OSC cannot be buried under our weather service. The White House and Congress should elevate the office by FY24. (OPRs: Department of Commerce (DOC), NSpC)

4. **DoD requires a process to rapidly acquire and constitute commercially-sourced capabilities for U.S. and allied warfighters.** Commercial space technology has forever changed the nature of conflict as evidenced by its contribution to the defense of Ukraine. Remote sensing, advanced analytics, and broadband communications are just a few of the tactical solutions that have had strategic impact on Russia's war in Ukraine. (OPR: DoD).

5. **In order to save the planet, we must get off-planet.** Advancements in off-world power production, manufacturing, and Lunar resource extraction will be foundational to the future multi-trillion dollar space economy. In order to lead, enabling strategy, policy, and law is required. Activities and human presence in space should be driven by an international

rules-based order and systems that uphold liberty and prosperity for all humankind. (OPR: NSpC).

6. **We must accelerate progress on strengthening the space industrial base.** The pace of progress towards past recommendations has not been brisk enough to lead. We must catalyze domestic manufacturing and supply chains to reduce costs and lead times while making headway on recommendations from previous years' SSIB recommendation.

Figure 3. View inside the Orbital Reef, a mixed-use space station in low Earth orbit for commerce, research, and tourism built as a joint collaboration between Blue Origin and Sierra Space. (Photo Credit: Sierra Space

# INTRODUCTION

*"We are in a race that we might win, but it is certainly conceivable that China may advance in certain areas of space faster than we do."*
— U.S. SENATOR JERRY MORAN[5]

At the conclusion of the 20th century, Americans identified scientific and technological advancements as the nation's greatest achievements, with the space program as the most-cited source of American pride.[6] The space program, and more specifically landing humans on the Moon, marked the culmination of a great power competition for space dominance with the former Soviet Union. The U.S. was able to triumph in the last space race because of a positive and unique relationship, a public private partnership, between the government and commercial industry. As a result of clarity of public purpose and private entrepreneurialism, the U.S. built a vibrant space ecosystem enabling it to be the first and only nation to send humans to the Moon, explore every planet in the Solar System, deliver free global positioning and timing, share Earth observation data globally for the benefit of humanity, create an international space station to facilitate global space development and engagement, achieve rocket reusability and ubiquitous commercial broadband communication. The world is freer and more prosperous for America having won the first space race against the first authoritarian space power, and having shaped an open system of commerce and information. Six decades later, U.S. leadership is once again contested by not one, but two authoritarian regimes, who seek to exert global dominance through control of the space domain.

Winning the New Space Race is a national imperative and a critical component of the preservation of liberty and prosperity in the 21st century for the United States, our allies, and partners. The rise of China as both an economic and space power is an imminent threat to democracy, the free market economy, and the international liberal order. China's many achievements in space are the result of Xi Jinping's "Space Dream" (航天梦) - a long term strategy that galvanizes a whole-of-nation approach toward a singular objective: displace the U.S. as the dominant space power both militarily and economically by 2045 (though some participants expressed concern that this may happen within the next decade absent significant actions).[7,8] In order to compete and retain its leadership role, the U.S. must clearly articulate a North Star vision for space that transcends administrations, and aggressively pursue a whole-of-nation action plan to achieve it. Both the vision and plan must integrate and synchronize efforts across civil, commercial and national security space and leverage both the disruptive innovation that is rapidly maturing within the new space economy domestically and abroad. They must fully enable the tools of commerce, innovation, and integration to achieve this outcome. The bold vision of the future and the incredible opportunities afforded by space demand it. The world order

---

[5] Bender, B. (2022). 'We're falling behind': 2022 seen as a pivotal lap in the space race with China. Politico.
[6] Pew Research (1999). Successes of the 20th Century. Pew Research.
[7] Pollpeter, K. et al. (2015). China Dream, Space Dream: China's Progress in Space Technologies and Implications for the United States. USCC.
[8] Pollpeter, K. (2021). China's Space Narrative: Examining the Portrayal of the US-China Space Relationship in Chinese Sources and its Implications for the United States. CNA

inherited by our children and grandchildren depends on it. A safe, secure, and sustainable space and Earth environment requires it. And the time for national unity of effort and purpose is now.

## PURPOSE

The State of the Space Industrial Base (SSIB) workshop is an annual meeting to assess progress and provide input and recommendations for the U.S. on its journey to secure the space future that honors commitments made in our Constitution to provide for the common defense, promote the general welfare, and secure the blessings of liberty to ourselves and our posterity. It solicits direct feedback from the U.S. space industrial base, investors, analysts, thought leaders and other stakeholders to assess our progress, suggest paths of synergy to build enduring national advantage, sustain and expand the space industrial base and broader national security innovation base. It is meant to provide actionable recommendations to actors in the space ecosystem enabled by leaders across the entirety of U.S. society.

## WORKSHOP OBJECTIVES

**Winning the New Space Race.** This workshop continues to assess the United States' progress toward preferred futures identified in *The Future of Space 2060 and Implications for U.S. Strategy*[9] which cautioned, "The U.S. must recognize that in 2060, space will be a major engine of national political, economic, and military power for whichever nations best organize and operate to exploit that potential."[10] The report defined the preferred future as one of thriving off-Earth human communities, a vastly expanded and self-sustaining economic sphere, with a balance of power that favored U.S. leadership in shaping a free and open system. As noted in the previous year, "what's at stake is no less than whether the largest geographic zone of human activity is one of democratic freedom and fair trade, or an autocratic exclusion zone."[11]

---

*"The space race of today is not like the one we faced with the Soviets. One can actually argue that the space race today has far more on the line. In this new era where the United States is being challenged across every sector, our American space enterprise cannot afford to lose focus or momentum. Therefore, to ensure the United States remains the global space power, we must commit to a set of pillars that are based on principles."*

- REP. ROBERT ADERHOLT, 2022[12]

---

**Continued Policy Momentum** - The nation made great progress with the successful continuance of the National Space Council, National Space Policy,[13] Artemis,[14] USSPACECOM, and U.S. Space Force across administrations and parties. Important steps have been taken with the first meeting of the

---

[9] Mozer, J. (2019). The Future Of Space 2060 & Implications For Us Strategy. DTIC.
[10] Air Force Space Command (2016). The Future of Space 2060 & Implications for U.S. Strategy.
[11] DIU (2021). State of the Space Industrial Base 2021 Infrastructure & Services for Economic Growth & National Security.
[12] Foust, J.(2022). House appropriator discusses space priorities. SpaceNews.
[13] White House (2020). National Space Policy. The White House.
[14] Howell, E. (2021). U.S. still committed to landing Artemis astronauts on the moon, White House says. Space.com.

expanded[15] National Space Council,[16] the release of the Space Priorities Framework,[17] the In-Space Servicing Assembly and Manufacturing Strategy,[18] and the recently released National Orbital Debris Implementation Plan.[19] Important actions were also taken to exercise interagency planetary defense actions,[20] and to create a White House-led Interagency working groups on debris[21] and Cislunar.[22] Important steps were taken in forwarding U.S. norms with the unilateral moratorium on destructive, direct-ascent anti-satellite (ASAT) missile testing,[23] release of the SECDEF's *Tenets of Responsible Behavior in Space*,[24] support for the United Nations' Open-Ended Working Group,[25] expansion of the Artemis Accords (doubling from 9[26] to 19[27]), release of the *Combined Space Operations Vision 2031*,[28] and inclusion of space in the Quad joint statement.[29] These efforts strengthen U.S. leadership at home and abroad, and help position the U.S. for the emerging area of strategic competition.

**The New Space Race Defined.** The U.S. is now in the dawn of a new era of strategic competition that threatens its peaceful and prosperous vision for the future on Earth and in space. China's recent emergence as a capable competitor in the space domain is concerning, as is its competing vision wherein the U.S. is permanently displaced as the leading economic and military power on Earth and beyond. China has defined both the timespan and scope of the new space race for sustainability, prosperity and the planet. China's national space timeline is 2045 where it intends to be the foremost spacepower: 1) eclipsing the U.S. in its economic activities and services; 2) eclipsing the U.S. in its foreign partner relations and shaping of the space governance system; 3) eclipsing the U.S. in feats that inspire the world, and 4) eclipsing the U.S. in military power. China's vision and goals are well laid out with top-level support from President Xi, General Secretary of the Communist Party of China. Chairman Xi has mobilized a whole-of-society effort towards his ends, embarking on a comprehensive range of ambitious goals, seeking to grow the market share and capabilities of its 'commercial' space sector with global finance,[30] and developing robust counter-space capabilities to make the U.S. uncomfortable. China has a clear vision for human expansion: it plans to build (in concert with Russia) a Moon base

---

[15] White House (2021) Executive Order on the National Space Council. The White House.
[16] NASA (2021). Vice President Highlights STEM in First National Space Council Meeting. NASA.
[17] White House (2021). United States Space Priorities Framework. The White House.
[18] White House (2022). In-space Servicing, Assembly, and Manufacturing National Strategy. The White House.
[19] White House (2022). National Orbital Debris Implementation Plan. The White House.
[20] NASA (2022). 2022 Interagency Tabletop Exercise.
[21] Office of Science and Technology Policy (2021). Federal Register :: Orbital Debris Research and Development Interagency Working Group Listening Sessions. Federal Register.
[22] Office of Science and Technology Policy (2022).Federal Register :: Request for Information; Cislunar Science and Technology Subcommittee. Federal Register.
[23] White House (2022). FACT SHEET: Vice President Harris Advances National Security Norms in Space.
[24] SECDEF (2021). Tenets of Responsible Behavior in Space.
[25] Acting Deputy Assistant Secretary of State Eric Desautels (2022). U.S. Statement to the Open Ended Working Group on Reducing Space Threats - U.S. Mission to International Organizations in Geneva. Department of State.
[26] Stimers, P. & Jammes, A. (2021).The Space Review: The Artemis Accords after one year of international progress. The Space Review.
[27] NASA (2022). NASA Artemis Accords.
[28] CSpO (2022). Combined Space Operations Vision 2031. Media.Defense.gov.
[29] White House (2022). Quad Joint Leaders' Statement. The White House.
[30] Hongpei, Z. (2022). China's commercial space sector picking up momentum, firms' financing expected to double in 2022. Global Times.

(International Lunar Research Station),[31] then to place humans on the Moon,[32] then to place humans on Mars.[33]

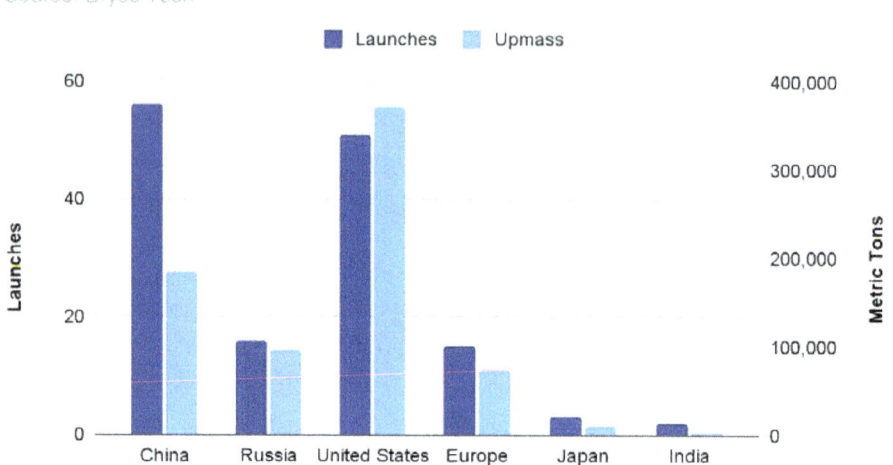

Figure 4. Space launches and total upmass by country. The combined activities of China and Russia pose a significant challenge to the U.S. and its spacefaring allies. (Credit: Bryce Tech)[34]

China intends to convert its entire fleet to reusable rockets – leveraging insights from leading innovators in the U.S. Space Industrial Base – and build next generation propulsion systems to enable space tourism and asteroid mining by 2040.[35] While China's official 5-year plans are more measured, prominent military and civilians space officials have articulated a more aggressive timeline to achieve commercial gigawatts of space-based solar power to dominate the future energy market, to use their Moon base to industrialize the Moon to build solar power satellites[36] and enable a $10 trillion/yr Moon-Earth Economic zone by 2050.[37] They are seeking international leadership in Planetary Defense [38] and International Lunar Search & Rescue.[39] They are using their space competencies to lure and lock in partners to their infrastructure, standards and governance systems.[40] Were China to win this race to industrialize the economic zone of space and secure its governance through leadership and might, it would have vast consequences for the security of the U.S. and its allies. The international order will have changed significantly.

---

[31] CNSA (2021). International Lunar Research Station (ILRS) Guide for Partnership.
[32] Freeman, M. (2020). China and the Moon. CASI.
[33] Reuters (2021). China plans its first crewed mission to Mars in 2033.
[34] U.S. data includes launches performed from New Zealand in support of U.S. customers.
[35] Goswami, N. & Garretson, P. (2020). Scramble for the Skies: The Great Power Competition to Control the Resources of Outer Space - 9781498583114. Rowman & Littlefield.
[36] Xinhua (2016). Exploiting earth-moon space: China's ambition after space station. China Daily
[37] Siqi, C. (2019). China mulls $10 trillion Earth-moon economic zone. Global Times.
[38] Jones, A. (2022). China is getting serious about planetary defense. Planetary Society; Young, C. (2022). China is building the world's most far-reaching radar system for planetary defense. Interesting Engineering; Young, C. (2022). https://interestingengineering.com/china-asteroid-deflection-mission. Interesting Engineering.
[39] IAAS (2022). The first international lunar search and rescue conference in Hainan (China).
[40] Per Jana Robinson, "As of May 2022, PSSI has identified globally 303 Chinese and Russian transactions targeting 83 countries...Out of the total number of recorded (bilateral) transactions China accounted for 147 transactions targeting 71 countries, Russia for 130 transactions targeting 43 countries."

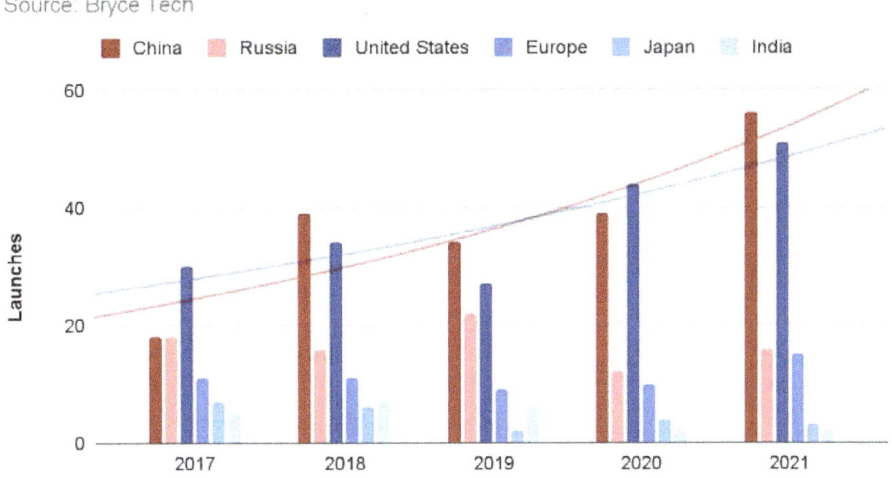

Figure 5. Trend in space launch cadence year-to-year by country. (Credit: DIU)[41]

---

*"The U.S. will most likely lose space superiority to China within the next decade."*
— ATLANTIC COUNCIL REPORT (2022)[42]

---

**Are We Racing Fast Enough?** The consensus of the workshop was that the U.S., as a whole-of-nation, is not competing with a collective sense of urgency. While there are isolated bright spots across NASA and DoD, the sense of urgency is not universally shared - especially within the vast bureaucracy that is constructively delaying U.S. commercial progress through regulatory burden.[43] As a result, our advantage continues to slip, with one prominent expert concluding that, absent serious changes, the U.S. may likely lose space superiority to China within the next decade, or by 2032.[44] Though subjective, polls taken by attendees indicated over the last year the U.S. had stood still, or regressed along *all* of the vectors of progress recommended last year.[45]

**Lack of Long-Term Vision & Policy Are the Bottleneck.** The key insight from the last two years remains the chief finding: the lack of a strategic-level guidance for a China-equivalent space vision for economic development and human expansion is causing all parts of the government and whole-of-nation to underperform. In a trickle-down effect, the absence of a clearly articulated national vision and guidance stunt the government's aspirations, leading to the preservation of status quo requirements and underperformance of government research and development.[46] In turn, the lack of ambitious demand signal from the government leads private industry to underperform its demonstrated potential.

---

[41] Data provided with permission of Rich Lesher, Bryce Tech.
[42] Eftimiades, N. (2022). Small Satellites: The Implications for National Security. Atlantic Council.
[43] Cahan, B. and Sadat, M. (2020). U.S. Space Policies for the New Space Age: Competing on the Final Economic Frontier. page 21.
[44] Eftimiades, N. (2022). Small Satellites: The Implications for National Security. Atlantic Council.
[45] See Appendix D.
[46] CED (2021). Back to Basic Research: An R&D Investment Plan to Enhance US Competitiveness. CED.

Workshop attendees identified a 'North Star Vision' as being both among the most important, and the easiest things the U.S. could achieve to make progress. Last year's report provided extensive guidelines for what the community would find acceptable and suitably ambitious to ignite national energies.[47] The consensus remains a vision of economic development and settlement -- at least on par with the expansive vision and timeline of our pacing competitor.

Figure 6. 2022 U.S. Space Industrial Base Scorecard (Source: USSF)[48]

**Subjective Evaluations.** We asked participants to provide a subjective assessment of the health of various parts of the space industrial sector. We asked them to rank each component on a 'stoplight' chart scale ranging from green (things are going well) to red (things are not going well), as well as to suggest the current trend compared to last year (see Figure 7). On average, participants were cautiously optimistic about the commercial sector, judging nearly all aspects -- investment, growth, innovation,

### SSIB'22 Workshop Participants Survey
Source: New Space New Mexico

| | | | | | |
|---|---|---|---|---|---|
| 3.5 | Investment | 3.3 | Atmospherics | 2.4 | Legislation |
| 3.5 | Growth | 3.2 | Markets | 3.5 | Accessibility |
| 2.9 | National Priorities | 3.1 | Executive Leadership | 3.4 | Universities |
| 4.0 | Innovation | 2.8 | Risk | 3.5 | New Starts |
| 3.8 | Technology | 2.3 | Incentives | 3.8 | Competition |
| 4.1 | Entrepreneurship | 2.6 | Cybersecurity | 2.5 | Human Capital |
| 2.6 | Strategy | 2.9 | Alliances | 2.9 | Enterprise View |
| | | 2.3 | National Supply Chain | 2.5 | Sustainability |

Figure 7. SSIB'22 workshop participants scorecard. (Credit: DIU)[49]

technology, entrepreneurship, atmospherics, newstarts and competition -- as yellow-green.[50] Most commercial categories also had favorable trends, except in atmospherics and markets. In contrast, participants were more pessimistic about human capital, sustainability, cyber security and national supply chain, rating them as a red-yellow and trending negatively. They were consistently pessimistic about those factors under government control, judging as yellow or yellow-red (and trending negatively since last year) several critical factors, including: strategy, legislation, executive leadership, incentives,

---

[47] DIU (2021). State of the Space Industrial Base 2021 Infrastructure & Services for Economic Growth & National Security.
[48] Olson, J. (2022). SSIB'22 Keynote Presentation: USSF Perspectives on Sustainability and Prosperity. USSF.
[49] Maethner, S. (2022). SSIB'22 Presentation: SSIB 2022 Survey Results. NSNM.
[50] This survey was taken a week prior to the Dow (INDU) dropping 876 points entering a Bull market on 13 June. Goldman, D. (2022). Dow tumbles 876 points and stocks enter bear market on worries of drastic rate hikes. CNN.

risk, and enterprise view. Interestingly, the greatest disparity between the government (USSF) and industry (workshop participants) assessments were in the areas of national priorities, strategy and alliances.

## GENERAL OBSERVATIONS

Major findings from the workshop include:

**China is determined and capable of winning the New Space Race** - China is making steady progress toward their goal of surpassing the U.S. as the dominant space power by 2045, with some participants assessing this may happen earlier, unless proactive measures are taken now to sustain our nation's leadership across all instruments of national power.

**New Space ecosystem at risk** - The agile engineering ecosystem that has become the hallmark of the new space era is at risk due to U.S. policy and procurement practices intended for a different era that embrace static requirements and slow (or stall) the fast-paced learning that occurs through rapid iteration. The U.S. bureaucracy is falling well short of implementing the national space strategy at a competitive pace.[51] Quite the contrary, their actions with regard to the space industry are impeding it.[52]

> *"The federal bureaucracy is a creature of Congress and the President. But agencies independently make policy and exert power: legislating by rulemaking; executing by implementation; and adjudicating by hearing complaints, prosecuting cases, and judging disputes."*
> — DAVID L. PALETZ, DIANA OWEN and TIMOTHY E. COOK[53]

**National Space Vision required** - An enduring North Star vision for America in space is an essential guidepost to remain competitive with a rapidly advancing China. Participants support 'Economic Development and Human Settlement' as that enduring vision to retain U.S. economic leadership, preserve the Earth, and protect other national interests.

**In order to protect the Planet, we must get off of the Planet** - Advancements in off-world power production, manufacturing and Lunar resource extraction will be foundational to shifting harmful greenhouse emissions from industrial activities beyond the Earth's biosphere and enabling the trillion-dollar space economy inherited by our future generations, starting with our children and grandchildren.[54]

**Commercial space technology has forever changed the nature of conflict** -- Space remote sensing, advanced analytics and broadband communications are tactical solutions that have delivered strategic impact as evidenced by their contribution to the defense of Ukraine.

---

[51] Berger, E. (2022). We got a leaked look at NASA's future Moon missions—and likely delays. ArsTechnica; Erwin, S. (2022). Satcom: Big money for military satellites, slow shift to commercial services. SpaceNews.
[52] Editorial Board (2022). SpaceX, Ocelots and the Mexican War. Wall Street Journal.
[53] D. L., Owen, D., & Cook, T. E. (2012). 21st century American government and politics. US: Creative Commons.
[54] O'Neill, G.K. (1974). The Colonization of Space. Physics Today, 27(9), 32-40.

## FOCUS

The 2022 State of the Space Industrial Base workshop focused primarily on one question, **"Are we making the progress we should be, and if not why not?"** For those who believe we are making progress, the question that follows is, **"Are we going fast enough?"** The consistent answer was "no," with the majority of participant respondents saying we were not moving with a sense of urgency. In response to **"What's the sense of urgency?,"** many experts including the Atlantic Council asserted that, if current trends continue, the U.S. may lose space superiority by 2032.

In assessing progress, this year's attendees -- members of the space industry, academia, and government -- reviewed and assessed the progress toward the findings and recommendations of the SSIB'21 report via survey (See [Appendix D](#)). Within working groups, they were also asked to evaluate and assess progress in terms of the following factors:

**Speed.** What must we do to accelerate change and achieve the vision in each of the working group areas? Which Space Futures demand the most urgent action? How are the threats from China and Russia impacting space and how should they be countered?

---

*"To explore the vast cosmos, develop the space industry and build China into a space power is our eternal dream."* – PRESIDENT XI JINPING[55]

---

**Commercial Success.** What must we do to allow the space industrial base to prosper and grow? In each area, how is commercial success affected by the different possible Space Futures? What government tools and resources can be brought to bear? How is the sourcing of components from China impacting the industry?

**Workforce.** In each area, how will the necessary talented and diverse workforce be developed, attracted to the space industry, and retained? How can we grow the space technology workforce through focused STEM education activities to enable greater adoption of commercial solutions enhancing civil and national security space? Should ITAR be relaxed to allow more contributions by foreigners?

**Policy.** What are implementable recommendations for space policy in each area that ensure the maintenance and expansion of the U.S. space industrial base to meet commercial, civil and defense needs of the United States, its allies and regional partners? How can USG practices regarding foreign ownership, control and influence (FOCI) be improved?

**Public-Private Coordination.** What implementable whole-of-government policies and actions are required to guarantee the U.S. space industrial base needed to ensure a space future supporting all instruments of U.S. national power? What are the first steps for the USSF to "buy what we can, build only what we must" toward the goal of increasing procurement of commercial products and services to 20% of Total Obligational Authority (TOA)?

**Manufacturing.** What short-term recommendations can help recover domestic manufacturing and grow the U.S. space industrial base in each working group's area?

---

[55] CNSA (2022). [China's Space Program: A 2021 Perspective](#). China National Space Administration.

**Infrastructure.** Will the U.S. lead in establishing a "space superhighway" and a Hybrid Space Architecture as the respective physical and digital infrastructure for an in-space logistics architecture that strengthens U.S. leadership in commercial, civil and national security space? All indications are that if Starship is successful, we'll be in a whole new era for space -- a truly transformative event.[56] Who is preparing for this and how? Who isn't?

**Sustainability.** How should space architectures pivot to maintain or increase the sustainability of space activities–by lowering costs, reducing or removing space debris, and making the industrial base more robust?

## MOTIVATIONS

**Space is an economic domain that is resource and opportunity rich** and provides many manufacturing environments that are too costly or impossible to reproduce on Earth such as ultra-high vacuum, super cold temperatures and microgravity.[57] There has been increasing investment in space companies, particularly startups. Yet, several challenges stand in the way of developing a truly robust space economy. Space as a domain currently lacks sufficient physical and digital infrastructure in the form of an integrated logistics and hybrid communications network architecture as a natural extension of the Earth's terrestrial global economy. Although the U.S. commercial space sector has experienced tremendous growth, U.S. supply chain challenges, inflation and depth of workforce issues threaten the economic viability of space domestically, as well as, the ability to maintain a strong national security space posture. Most government space projects are independently developed and funded in organizational silos, and there is little desire or incentive to coordinate between them. Financial tools are lacking which perpetuates an investment climate that benefits only a few, rather than for all. The rate of obsolescence in legacy space systems is accelerating with little tangible indication of change.[59] In the absence of commercial innovation and agile production capability, most systems will reach obsolescence within five years.

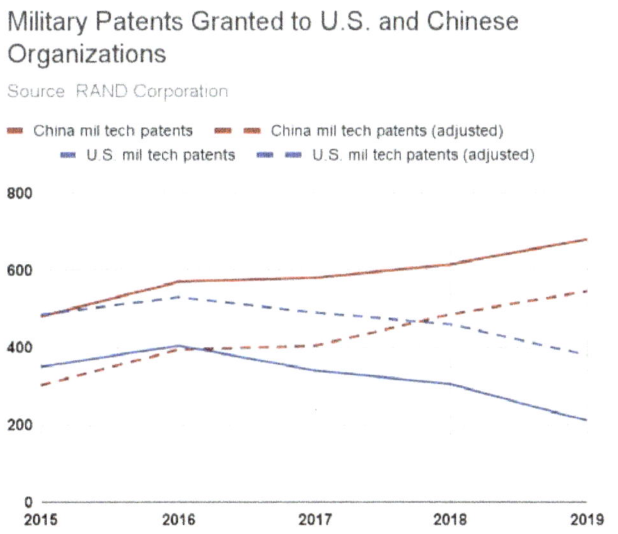

Figure 8. Raw and Quality-adjusted Military Patents Granted to U.S. and Chinese Organizations (Credit: RAND)[58]

---

[56] Simberg, R. (2021). Walmart, But for Space. The New Atlantis.
[57] The White House Office of Science and Technology Policy hosted a microgravity R&D workshop in late July to continue the conversation about the future of sustainable research platform access.
[58] Weinbaum, C. et al (2022). Assessing Systemic Strengths and Vulnerabilities of China's Defense Industrial Base. RAND Project Air Force.
[59] Davis, J., Mayberry, J. & Penn, J. (2019). On-orbit servicing: inspection, repair, refuel, upgrade and assembly of satellites in space. Aerospace Corp.

**China continues making steady gains in innovation** - Although many U.S. companies choose to keep their intellectual property internal and not file patents, it is nevertheless concerning that China is making steady gains in patent quality. As assessed by RAND in Figure 8 above, China has made steady gains in Quality-Adjusted Military Patents, surpassing the U.S. since 2018.[60] China is now assessed by the former Deputy Assistant Secretary of the Air Force for Acquisition Contracting, Maj. Gen. Cameron Holt, to be fielding new systems "five to six times" faster than the U.S.[61] Holt also noted, "In purchasing power parity, they spend about one dollar to our 20 dollars to get to the same capability," and "We are going to lose if we can't figure out how to drop the cost and increase the speed in our defense supply chains."[62]

At the same time, space is an area of global competition with increasing potential for conflict as more nations, and non-state actors, achieve spacefaring status. With regard to national defense, a strong U.S. space industrial base is essential to:

- Provide new space capabilities that leverage state-of-the-art hardware and software.
- Ensure the ability to defend increasingly critical civil and commercial space capabilities during peacetime and conflict.
- Improve resilience by increasing the availability and survivability of systems, components and microelectronics that remain tightly coupled to the current state of advanced technology.
- Leverage cost savings through the procurement of commercial solutions in order to maximize investment in bespoke government solutions critical to national security.
- Maintain and expand U.S. space capabilities as an element of overall national power and security.
- Demonstrate the feasibility of new, disruptive technologies that will change the status quo.

**Space will provide the greatest economic opportunity, but with significant first mover advantage** - There is a growing set of analyses that project space as an increasingly relevant domain of both robotic and human activity. Space has become a rapidly expanding source of commercial, civil and military national power. This, in turn, puts U.S. capabilities at increased risk as peer competitors seek to displace our leadership and deny this advantage. U.S. space power will be required to adapt and grow as needed to protect U.S. national interests and liberal democracy globally. Most importantly, participants emphasized that:

- The U.S. cannot fast follow China.
- The U.S. cannot become China to beat China.
- The U.S. must take a whole-of-nation approach, assembling the best talent from industry, government and academia to leapfrog China.
- The U.S. must be more inclusive and work collaboratively with the commercial, civil and national security space innovators of its allies and partners around the world.
- The key U.S. government space actors, USSF and NASA, must collaborate to enable many of the advanced technologies that will assure continued U.S. space leadership.

---

[60] Weinbaum, C. et al (2022). *Assessing Systemic Strengths and Vulnerabilities of China's Defense Industrial Base*. RAND Project Air Force.
[61] Nedwick, T. (2022). *China Acquiring New Weapons Five Times Faster Than U.S. Warns Top Official*. The Drive.
[62] Nedwick, T. (2022). *China Acquiring New Weapons Five Times Faster Than U.S. Warns Top Official*. The Drive.

The Department of Defense has identified critical technology areas requiring accelerated space industrial base innovation and the advancement of technology readiness. These areas are described in the United States Space Priorities Framework,[63] U.S. Space Science and Technology Strategy[64] and the USSF Space Futures II workshop report. In addition, the U.S. Congress specifies critical space technologies in the National Defense Authorization Act.[65]

Figure 9. A USAF Pararescue team working with 45th Operations Group Detachment 3 personnel and mission partners on tactics and procedures for astronaut rescue and recovery operations. (Credit: USSF)

**An American Comeback is Possible** - Despite China's recent gains, several long-term trends argue a U.S. comeback is possible. This year, the U.S. economy is forecast to grow at a faster rate than China's for the first time since 1976.[66] China's national security innovation base relies on aging, foreign-trained talent and faces shortfalls in workers with needed skills.[67] China is projected to reach its Lewis Turning Point[68] within the next decade.[69] The growing expectations among young workers have given rise to a "lay flat" (躺平, Tǎng Píng) movement among middle-class office workers looking to provide minimalist input into unfulfilling work; a third of recent graduates (working in China) quit their first job within six months of graduation due to dissatisfaction with their duties, work-life balance, or compensation.[70] China will start to lose manufacturing jobs to countries like India unless it can

---

[63] White House (2021). United States Space Priorities Framework.
[64] DoD (2015). Department Of Defense Space Science And Technology Strategy 2015. DTIC; a 2022 version is complete and forthcoming.
[65] The FY22 NDAA included funding for Space Range of the Future, Hybrid Space Architecture and an Accelerated Cislunar Flight Experiment.
[66] Morgan, L. (2022). China's Great Gamble: A Conversation with Barry Naughton. IGCC.
[67] Bond, M.S. (2022). China's Defense Industrial Base: Labor Shortage and Skill Gaps Will Continue to Hamper Progress PAF-1P-747. RAND Project Air Force.
[68] The Lewis turning point is a situation in economic development where surplus rural labor is fully absorbed into the manufacturing sector. This typically causes agricultural and unskilled industrial real wages to rise. (Source: Wikipedia).
[69] Bond, M.S. (2022). China's Defense Industrial Base: Labor Shortage and Skill Gaps Will Continue to Hamper Progress PAF-1P-747. RAND Project Air Force.
[70] Bond, M.S. (2022). China's Defense Industrial Base: Labor Shortage and Skill Gaps Will Continue to Hamper Progress PAF-1P-747. RAND Project Air Force.

increase its own labor pool.[71] But increasing its labor pool is unlikely since China's population is on track to turn down this year for the first time since the great famine of 1959-1961 -- a decade sooner than expected -- and China's population is forecast to decrease to 587 million in 2100, less than half of what it is today.[72]

## BARRIERS

In the absence of a clearly articulated North Star vision that focuses efforts, streamlines bureaucratic processes, and energizes the public sector, the strategic end state of the U.S. retaining its leadership role in space is unwittingly compromised. Examples of barriers impeding the United States' ability to compete in the 21st Century provided by the SSIB Workshop participants include:

**Bureaucratic Delays** - Many advanced U.S. commercial launch vehicles and satellites are sitting idle in high bays and warehouses, often for months, if not years, while disparate federal agencies mull over impact studies, environmental assessments, spectrum licenses, launch licenses, export control, shutter control, and a seemingly endless list of other requirements before an approval is issued to operate.

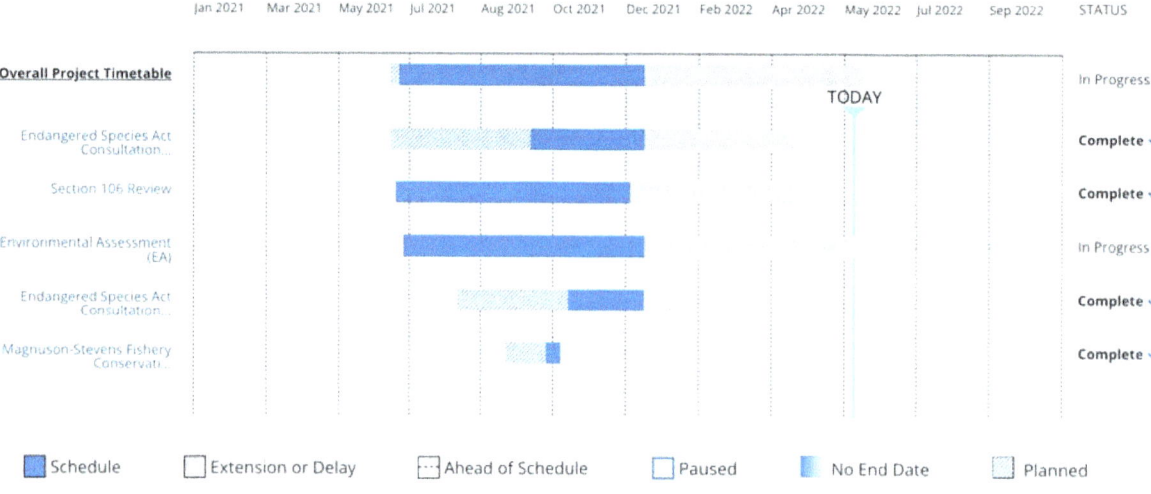

Figure 10. FAA Starship Approval Timeline (as of 6 Jun 2022)[73]

**Quest for Standards** - A number of government- and industry-sponsored efforts (i.e., OSAM, CONFERS) openly advocate for formal establishment of standards across the new space ecosystem. Yet, some, including the National Institute of Standards and Technology (NIST), caution that establishing regulatory standards too early tends to stifle innovation rather than promote or accelerate it.[74]

**Requirements-based acquisition culture** - Favors bespoke, government-defined solutions that often limit the scope of full and open competition and takes years to deliver impact to end users. Bespoke solutions tend to be more exquisite and more costly to design, build and sustain.

---

[71] Bond, M.S. (2022). China's Defense Industrial Base: Labor Shortage and Skill Gaps Will Continue to Hamper Progress PAF-1P-747. RAND Project Air Force.
[72] Peng, X. (2022). China's population is about to shrink, here's what it means for the world. The Conversation.
[73] FAA (2022). Permitting Dashboard: SpaceX Starship/Super Heavy Launch Vehicle Program at the SpaceX Boca Chica Launch Site in Cameron County, Texas | Permitting Dashboard.
[74] Tassey, G. (1999). Standardization in Technology-Based Markets. NIST.

**Overclassification** - There are more opportunities to advance U.S. interests by releasing data to additional stakeholders. Strategic competition requires space capabilities to be "allied by design." Overclassification stymies trust and inhibits the collaboration and trust required between allies and partners to compete in the 21st century.[75]

**The U.S. budgeting process** - The Planning, Programming, and Budgeting System (PPBS) process established in 1962 assumes an environment that no longer exists -- where the U.S. Government is primal in research & development, and largely uncontested across all domains including space.[76] The PPBS predates the global digital revolution and accommodates neither the procurement of agile software or commercial services as 'real things' that are separate and distinct from hardware or materiel. Recognizing the need for reform of this antiquated system, in Public Law 117-81, Congress created a 14-member independent commission within the legislative branch to reform the DoD Planning, Programming, Budgeting and Execution system.[77]

**Programs of Record** - or "line item of record" is a highly structured procurement in the Future Years Defense Program. Service Secretaries lack flexibility to manage siloed appropriations and 'colors of money' independently of major programs, or during the Continuing Resolutions (see Figure 11), thus delaying execution.

**Export laws** - Six years ago, the late Senator John McCain championed changes necessary to encourage technology transfer and increase industrial cooperation with U.S. allies;[78] and yet updating our nation's export controls process is long overdue.

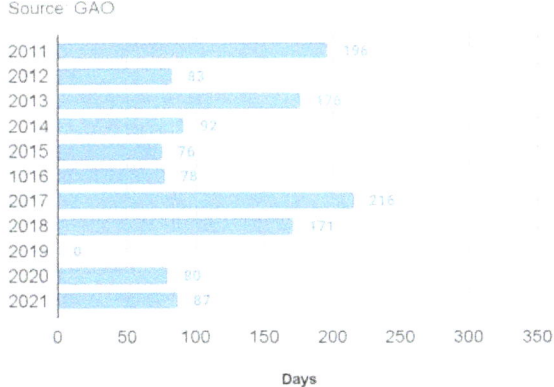

Figure 11. A total of 3.4 years have been lost since 2011 due to delayed execution of defense budgets (Credit: GAO)[79]

**Policy Adherence** - The U.S. released the State of Competition in the Defense Industrial Base Report in February of this year.[80] Timely adoption of its five recommendations is paramount.

---

*"The Chinese government steals staggering volumes of information and causes deep, job-destroying damage across a wide range of industries - so much so that, as you heard, we're constantly opening new cases to counter their intelligence operations, about every 12 hours or so."*

— CHRISTOPHER WRAY, Director FBI, 2022[81]

---

[75] Carberry, S. (2022). Over-classification, lack of standards stymies Allied Space Forces. National Defense Magazine.
[76] Whitley, J. and Pejic, G. (2021). Senate Commission to fix Defense budgeting is right on the mark. War on the Rocks.
[77] DoD (2022). DoD Announces Appointments to the Commission on Planning, Programming, Budgeting and Execution Reform.
[78] Greenwalt, B. (2022). The NTIB is dying; Is AUKUS. next? Congress must apply life support soon. Breaking Defense.
[79] GAO (2021). DoD Has Adopted Practices to Manage within the Constraints of Continuing Resolutions. GAO-21-541
[80] U.S. (2022). State of Competition in the Defense Industrial Base. Defense Media.
[81] FBI (2022). Remarks by Christopher Wray: Countering Threats Posed by the Chinese Government Inside the U.S.

**Competitors' insatiable meddling with U.S. Industry** - Intellectual Property (IP) theft, corporate espionage and problematic FOCI continue somewhat unabated.

**Lack of financial engineering to shape the expanding space economy** - In the absence of Wall Street leadership, other nations are strengthening and diversifying finances of the new space economy.

**Investor timidity** - Special Purpose Acquisition Company (SPAC)-facilitated transactions should not be the only means for an enthusiastic public to invest in the new space economy. Financial engineering is required to enable diversified instruments to grow and sustain investor interest.

**Competition for workforce** - Newly graduated aerospace and electrical engineers are accepting work in other sectors of the economy where they see better pay, benefits and job security.[82]

## OPPORTUNITIES

**Transformation of Space Operations** - We are at the dawn of a new era marked by increased viability of reusable launch vehicles and sustainable satellites that will be modular, refuelable and maneuverable.

**Cloud & Internet in Space** - The most interesting things we will do digitally in space will require access to the cloud and the internet - both are extremely limited off the Earth's surface.

**Hybrid Space Architecture** - Seeks to assure information superiority by connecting commercial, civil and national security systems together in space using a secured, low latency, multi-path communications architecture upon which the cloud is foundational. Includes both physical (satellites) and digital (software-defined) infrastructure necessary to set the first cornerstone for the 'Outernet' - the internet in space.

**In-space Logistics Architecture** - The physical infrastructure that, when paired with Hybrid Space Architecture as a digital and physical infrastructure, serves as the backbone of the new space economy upon which many other businesses will be built. In-Space Servicing and Assembly (ISAM) and Space Mobility and Logistics (SML) are vital enablers.

Figure 12. Astranis engineers complete acceptance testing of 'Arcturus,' the first MicroGEO communications satellite, in preparation for launch in 2022. (Source: Astranis)

**Advanced Propulsion** - Direct flight to the Moon in hours not days, and Mars in weeks not months will become possible with advanced nuclear propulsion technology in development today.[83]

---

[82] Jackson, A. (2022). Wall Street interns are making 30% more this year - some making over $10,000 a month. CNBC.
[83] Conversation with DARPA.

**Limitless Green Energy** - Harnessing the Sun's energy above the atmosphere and safely transmitting it to the Earth's surface is critical to evolving away from a carbon economy.

**Alliances and Partnerships** - The Great Autocracies (China, Russia, etc.) become much smaller problems when the U.S. stands together in unity with allies and regional partners who share a common vision and belief system.

---

*"The advantage America enjoys today stems from our space industrial base. We must work together to ensure that it remains a strong, effective, and innovative partner in sustaining American space superiority. The 2020 Defense Space Strategy recognizes that commercial space activities have expanded significantly in both volume and diversity."*

– GEN JOHN W. RAYMOND, First Chief of Space Operations[84]

---

## WORKING GROUP BREAKOUTS

The State of the Space Industrial Base 2022 Workshop was organized with specific breakout sessions enabling dialogue between stakeholders from industry, academia and government to address the factors above and answer the most important question, "Are we making the progress we should be, and if not why not?" Chatham House Rules were observed within the breakout sessions to facilitate a more comprehensive discussion. Specific working group breakouts for SSIB'22 included:

- Launch Services
- Hybrid Space Communications
- In-Space Transportation & Logistics
- Next Generation Power & Propulsion
- Remote Sensing & Traffic Management
- Policy & Finance
- Workforce & Science, Technology, Engineering and Math (STEM) Education

## OVERALL ASSESSMENT SINCE SSIB'21

U.S. leadership is strong in terms of the evolution of new space policy and strengthening space budgets. The challenge is in the implementation of important and enabling changes needed to compete in the new strategic environment. Overall, workshop attendees surveyed during SSIB'22 responded that insufficient progress had been made on all but a few recommendations. An expanded discussion on the 17 recommendations from SSIB'21 are detailed in Appendix C.

---

[84] Olson, J. (2022). SSIB'22 Keynote Presentation: USSF Perspectives on Sustainability and Prosperity. USSF.

## KEY ACTIONS & RECOMMENDATIONS

From the above observations and additional inputs from the working groups, participants advocated for or expressed interest in six overarching recommendations for action, which reflect the numerous inputs from workshop participants and their best assessment regarding which agency(cies) are in the best position to forward the necessary change:

1. **Establish an enduring U.S. North Star Vision for Space** as an essential guidepost to sustain the United States' competitive advantage against a rapidly advancing China that is now collaborating with Russia. This whole-of-nation vision must be as clear and ambitious in scale and timeline as the PRC and more inclusive of international collaboration across the spectrum of commercial, civil and national security space activities. (OPRs: National Security Council (NSC), National Space Council (NSpC), Office of Science and Technology Policy (OSTP))

2. **Enable National Security Priority Processing of Licenses/Environmental Clearances for Critical Space Systems**. The agile engineering ecosystem that has become the hallmark of the new space era is at risk due to U.S. policy and procurement practices put in place by Robert McNamara in 1962 when the government was the source of innovation leadership. Today, we must use Defense Production Act (DPA) Title III and other authorities such as the Defense Priorities and Allocation System (DPAS) to reduce bureaucratic delay on critical needs for national security. (OPRs: NSC, NSpC, OSTP, Office of the Secretary of Defense (OSD))

3. **Elevate the Office of Space Commerce (OSC) to report directly to the Secretary of Commerce**. In a whole-of-nation strategy where commerce and economic growth are central to strategic competition, the OSC cannot be buried under our weather service. The White House and Congress should elevate the office by FY24. (OPRs: Department of Commerce (DOC), NSpC)

4. **DoD requires a process to rapidly acquire and constitute commercially-sourced capabilities for U.S. and allied warfighters.** Commercial space technology has forever changed the nature of conflict as evidenced by its contribution to the defense of Ukraine. Remote sensing, advanced analytics, and broadband communications are just a few of the tactical solutions that have had strategic impact on Russia's war in Ukraine. (OPR: DoD).

5. **In order to save the planet, we must get off-planet.** Advancements in off-world power production, manufacturing, and Lunar resource extraction will be foundational to the future multi-trillion dollar space economy. In order to lead, enabling strategy, policy, and law is required. Activities and human presence in space should be driven by an international rules-based order and systems that uphold liberty and prosperity for all humankind. (OPR: NSpC).

6. **We must accelerate progress on strengthening the space industrial base.** The pace of progress towards past recommendations has not been brisk enough to lead. We must catalyze domestic manufacturing and supply chains to reduce costs and lead times while making headway on recommendations from previous years' SSIB recommendations.

Figure 13. Earth and Dragon's nose cone (Credit: Inspiration4 crew)

*"This tiny Earth is not humanity's prison, is not a closed and dwindling resource, but is in fact only part of a vast system rich in opportunities, a High Frontier which irresistibly beckons and challenges the American genius."*

- HOUSE CONCURRENT RESOLUTION 451, 1978[85]

---

[85] Teague, O. (1978). Excerpt from House Concurrent Resolution introduced in 1978 directing the Congress and executive agencies to determine how they may aid in achieving national goals in outer space. Library of Congress.

THE WHITE HOUSE
WASHINGTON

CONFIDENTIAL                                April 11, 1962

NATIONAL SECURITY ACTION MEMORANDUM NO. 144

TO:     The Vice President, (as Chairman, National
        Aeronautics and Space Council)
        The Secretary of Defense
        The Secretary of Commerce
        Administrator, National Aeronautics and Space Agency
        Director, Bureau of the Budget
        Director, Office of Emergency Planning

SUBJECT: Assignment of Highest National Priority to
         the APOLLO Manned Lunar Landing Program

In response to a recommendation by the National Aeronautics and Space Council, which approved a proposal by the Administrator, National Aeronautics and Space Agency, the President under the authority granted by the Defense Production Act of 1950 today established the program listed below as being in the highest national priority category for research and development and for achieving operational capability:

    APOLLO (manned lunar landing program, including
    essential spacecraft, launch vehicles, and facilities).

                                        McGeorge Bundy

Information Copy to:
    General Maxwell Taylor              cc: Mrs. Lincoln
                                            Mr. Bundy File (3)
                                            C. Johnson
                                            NSC Files (2)

DECLASSIFIED
NSC ltr. dated April 9, 1979
By MJM   NARS, Date 5/1/79

Dispatched 4/12/62

Figure 14. Implementation of a North Star vision to win the first space race. A declassified National Security Action Memoranda, NASM-144, assigning the highest priority to the Apollo Manned Lunar Landing Program in 1962. (Source: John F. Kennedy Presidential Library & Museum)[86]

---

[86] Digital identifier: JFKNSF-335-024

# THE NORTH STAR VISION: ECONOMIC DEVELOPMENT & HUMAN SETTLEMENT OF SPACE

*"Exploration by a few is not the grandest achievement. Occupation by many is grander."*
—DR. JOHN H. MARBURGER III, former OSTP Director [87]

## BACKGROUND

**Reinforcing Previous Recommendations** - The top recommendation for the third consecutive year remains the lack of clarity with regard to a whole-of-nation vision, or end state in the 21st century, that is both enduring and inspirational. Human settlement in Space is the logical choice as it enables so much more than simply exploration and national security. America's grand strategy for space must inspire all Americans including those who currently don't see their future role in space. If humans live in space, then they require healthcare workers, engineers, miners, metal workers and Amazon delivery drivers (or their equivalents). Space will become a natural extension of the terrestrial economy once humans are permanently residing in it. A North Star vision for space must make sense, endure generations and guarantee the preservation of liberty and prosperity in a world where the opinion of our grandchildren matters. Most importantly, it must provide an alternative to China's grand strategy or 'Dream'.

**Why This Is Important Now** - A national vision serves multiple objectives.[88] It reflects key strategic choices statespersons have made about opportunities and threats. It focuses federal agencies on the objectives and activities which critically contribute to the success of U.S. foreign policy, thereby directing effort and establishing criteria. Since a North Star vision is public, it sends diplomatic messages that reassure allies, attract partners, and put rivals on notice the U.S. intends to contest their leadership and deter possible aggression. It speaks directly to the society that must provide the key contributions of people and money upon which it will be executed, and clearly establishes a link between the activities of subordinate organizations and national purpose. It provides a shared conceptual framework from which to organize coordination, cooperation, reduces the fog and friction of uncertainty, enabling harmony and cohesion in execution across diverse elements. It supplies a compelling narrative to motivate individual members to work for national purposes whether in government or industry. And it

> *"The U.S. must develop and execute a grand strategy for space recognizing space's importance and enhancing our advantages. This strategy must encompass the near-term future, with space oriented as a source for augmenting terrestrial power, and the long-term future, encompassing space across the Cislunar expanse and beyond as a domain in itself for human action."*
> – USSF SPACE FUTURES WORKSHOP REPORT, 2021

---

[87] Marburger, J. (2008). 2008 Goddard Memorial Symposium speech (Dr. John Marburger was Director, Office of Science and Technology Policy under President George W. Bush).
[88] Posen, B.R. (2016). Foreword: Military doctrine and the management of uncertainty. Journal of Strategic Studies.

provides the soft-power of inspiration, confidence and aspiration by which the global audience measures and develops sympathy for and conveys legitimacy upon U.S. leadership. It stimulates the imagination and consciousness of our children to pursue careers in Science, Technology, Engineering, the Arts and Math (STEAM) in order to carry this grand vision forward for the benefit of future generations. Thus, it is the best path to ensure humanity doesn't become an endangered species.

**Is there a new Space Race?** Yes, although it is somewhat unacknowledged by the U.S. In action, many are oblivious to it, or worse, ignore it. As the victors of the last space race with the Soviet Union in the 1960's, history reminds us that we came perilously close to losing that race. As our capabilities grew, we readily dismissed strategic competition from abroad including China and to a lesser extent Russia today.

> *"We are witnessing a second space race. This time the primary components are not status-seeking for a global audience, but lasting economic power. It is the beginning stage of a competition to secure positional, logistical, and industrial advantage. It is a competition to secure the material wealth of space resources, and to convert that wealth into power to shape the international system."*
>
> -- NAMRATA GOSWAMI & PETER GARRETSON, 2020[89]

For decades, the U.S. helped foster an institutionalized view of spaceflight as the playground of elite nation-states. We must galvanize a strategic vision that bridges all instruments of national power in pursuit of a common goal. In the absence of a North Star vision, there is division, duplication and a growing sense of fait accompli.[90]

**What is the Space Race?** - The competition represents a major inflection point not just for the 21st Century but for all of human history. The New Space Race seeks to achieve nothing less than the permanent establishment of the first off-planet, human settlement propelled and sustained by a thriving to-, in-, and from-space economy. It will be forged from a Cislunar industry that knits together the physical, digital and power infrastructure required to support the future needs of Earth, and to extend human activity to Mars and beyond. Such a vision calls for international collaboration towards a Free and Open Cislunar structure much like the Administration's Open Indo-Pacific[91] strategy to counter China's 'Debt Trap' diplomacy enabled by the Belt and Road

Figure 15. Taikonaut on China's Space Station (Credit: Licensed from Getty Images)[92]

Initiative.[93] A compelling, comprehensive, and clearly articulated North Star vision, with timelines to ensure the U.S. remains an undisputed space leader, must touch all instruments of national power -

---

[89] Goswani, N. & Garretson, P. (2020). *Scramble for the Skies: The Great Power Competition to Control the Resources of Outer Space*. Lexington Books.
[90] Bender, B. (2021). 'We're falling behind': 2022 seen as a pivotal lap in the space race with China. Politico.
[91] Blinken, A. (2021). A Free and Open Indo-Pacific. Speech to Universitas Indonesia. U.S. Department of State.
[92] Jennings, J. (2021). In China-US Space Race, Beijing Uses Space Diplomacy. VOA.
[93] Tyson, A.S. (2021). Is China ensnaring poor countries by building their infrastructure? Christian Sci. Monitor.

Economic/Industrial, Coalitional, Military, Informational, Technological - and be woven into similar terrestrial strategies promoting liberty and prosperity.

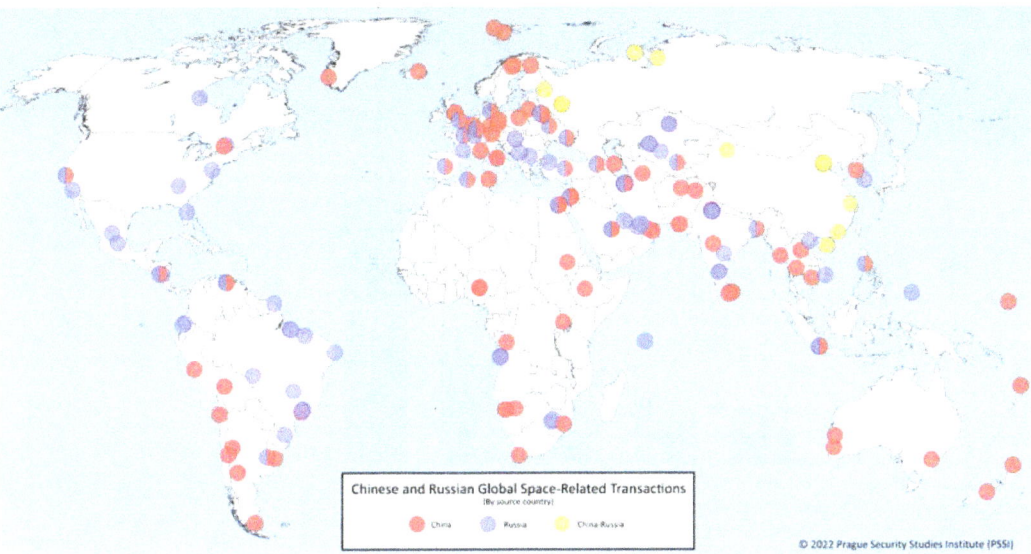

Figure 16. Chinese and Russian Global Space Related Transactions. (Credit: PSSI)[94]

**The Ground-Based Space Race** - America must be aware that the autocratic space powers are bringing a strong ground-component to the new space race, aggressively seeking space partnership agreements on planet Earth. As of May 2022, the think tank PSSI identified globally 303 Chinese and Russian transactions targeting 83 countries (see Figure 16). Fourteen of those transactions are between China and Russia. Out of the total number of recorded bilateral transactions, China accounted for 147 transactions targeting 71 countries. Russia for 130 transactions targeting 43 countries. Twelve additional international multilateral agreements were sponsored by Russia and/or China.[95]

## CURRENT STATE

**Space Industrial Base** - 2021 was a record-breaking year for the U.S. Space Industrial Base. According to BryceTech, a total of $15.4 billion was invested across 241 start-up space ventures during 2021, smashing the previous record of $7.7 billion invested during 2020 (see Figure 17).[96] Mergers, acquisition and initial public offerings (IPOs) were equally unsurpassed. However, while the United States space industrial base is on an upward trajectory, participants expressed concerns that the upward trajectory of the PRC is even steeper, with a significant rate of overtake, requiring urgent action. The fundamental tonic is to mobilize still greater energies with an enlarged vision and broader set of policy and finance enablers.

**Underinvestment in Supply Chain** - Despite the impressive growth in the sector, U.S. supply chains have failed to match the growing demand of a burgeoning commercial space industrial base, driving

---

[94] Robinson, J. (2022). Presentation at the Prague Space Security Conference.
[95] Robinson, J. (2022). Presentation at the Prague Space Security Conference.
[96] Bryce Tech (2022). Start-Up Space 2022: Update on Investment in Commercial Space Ventures. Bryce Tech.

further dependencies on foreign sources for critical parts and components.[97] Logistics challenges of 2021 translated into longer lead times for critical parts and significantly increased cost for parts.

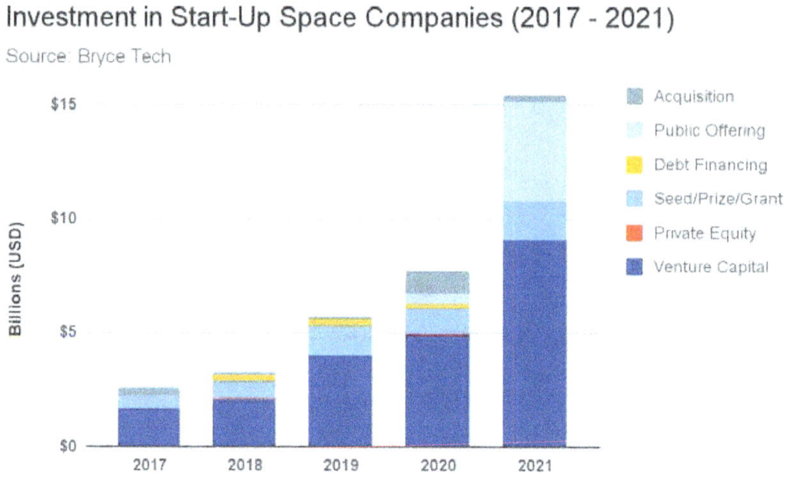

**Investment in Start-Up Space Companies (2017 - 2021)**
Source: Bryce Tech

Figure 17. Investments in space start-up companies nearly doubled in 2021. (Credit: DIU)

**Lack of Production Contracts** - Private investors voiced their frustration that despite the significant investments made in technologies critical to national security and the economy, the U.S. Government has been sporadic in fulfilling the role of an early-adopter customer in awarding meaningful production contracts for products and services to address the demand signal being sent to Silicon Valley and other innovation centers.[98]

> *"Time is running out... We have, at most, two years before founders walk away and private capital dries up. And many, many startups will go out of business waiting for DoD to award real production contracts."*
>
> – KATHERINE BOYLE, GP, Andreessen Horowitz[99]

**Policy** - The nation made great progress with the successful continuance of the National Space Council, National Space Policy,[100] Artemis,[101] USSPACECOM, and U.S. Space Force across administrations and parties. Important steps have been taken with the first meeting of the expanded[102] National Space Council,[103] the release of the Space Priorities Framework,[104] the In-Space Servicing Assembly and Manufacturing Strategy,[105] and the recently released National Orbital Debris Implementation Plan.[106] Important actions were also taken to exercise interagency planetary defense

---

[97] Foust, J. (2021). Space industry feels varying effects of supply chain disruptions. SpaceNews.
[98] Insinna, V. (2021). Silicon Valley warns the Pentagon: 'Time is running out'. Breaking Defense.
[99] Boyle, K. (2021). Status update on Reagan Defense Forum. Twitter.
[100] White House (2020). National Space Policy. The White House.
[101] Howell, E. (2021). U.S. still committed to landing Artemis astronauts on the moon, White House says. Space.com.
[102] White House (2021) Executive Order on the National Space Council. The White House.
[103] NASA (2021). Vice President Highlights STEM in First National Space Council Meeting. NASA.
[104] White House (2021). United States Space Priorities Framework. The White House.
[105] White House (2022). In-space Servicing, Assembly, and Manufacturing National Strategy. The White House.
[106] White House (2022). National Orbital Debris Implementation Plan. The White House.

actions,[107] and to create a White House-led Interagency working groups on debris[108] and Cislunar.[109] Important steps were taken in forwarding U.S. norms with the unilateral moratorium on destructive, direct-ascent anti-satellite (ASAT) missile testing,[110] release of the SECDEF's *Tenets of Responsible Behavior in Space*,[111] support for the United Nations' Open-Ended Working Group,[112] expansion of the Artemis Accords (doubling from 9[113] to 19[114]), release of the *Combined Space Operations Vision 2031*,[115] and inclusion of space in the Quad joint statement.[116]

**Budget** - The Administration, in its FY23 budget request also proposed significant increases in funding for NASA (+$1.93 billion, or +8%),[117] Space Force (+$6.45B or +36%),[118] and the Office of Space Commerce (+$78M),[119] and appropriators may approve an even bigger budget for the office of Space Commerce ($87.7 million, up from $16 M currently) as well as the request to move it out of NESDIS to report directly to the NOAA Administrator.[120]

> *"A U.S. North Star vision for space must focus on economic development and prosperity with the goal of an in-space human-centric economy."* – SSIB'22 Participants

**It's Still Not Enough**. While the above are important actions, workshop participants still assessed that U.S. policy was underperforming in supplying the necessary vision and framework, especially when compared with China's Dream, and that the absence of such a North Star Vision constituted the major bottleneck in maintaining enduring U.S. advantage. There is a need to cast a broader, more inspirational and inclusive vision for all Americans. There is consensus that it must focus on economic development and prosperity with the goal of an in-space human-centric economy. The Space Priorities Framework is an important first step, but a stronger connection between space and America's more important challenges, a more inspirational quality, and a set of ambitious goals or timelines would further strengthen the vision. The document thus does not adequately provide a vision and compelling narrative to repair eroding U.S. advantage and secure economic prosperity, nor provide the necessary focus and urgency to keep to deadlines.[121] Despite the narrative examples set by China, the UK,[122]

---

[107] NASA (2022). 2022 Interagency Tabletop Exercise.
[108] Office of Science and Technology Policy (2021). Federal Register :: Orbital Debris Research and Development Interagency Working Group Listening Sessions. Federal Register.
[109] Office of Science and Technology Policy (2022).Federal Register :: Request for Information; Cislunar Science and Technology Subcommittee. Federal Register.
[110] White House (2022). FACT SHEET: Vice President Harris Advances National Security Norms in Space.
[111] SECDEF (2021). Tenets of Responsible Behavior in Space.
[112] Acting Deputy Assistant Secretary of State Eric Desautels (2022). U.S. Statement to the Open Ended Working Group on Reducing Space Threats - U.S. Mission to International Organizations in Geneva. Department of State.
[113] Stimers, P. & Jammes, A. (2021).The Space Review: The Artemis Accords after one year of international progress. The Space Review.
[114] NASA (2022). NASA Artemis Accords.
[115] CSpO (2022). Combined Space Operations Vision 2031. Media.Defense.gov.
[116] White House (2022). Quad Joint Leaders' Statement. The White House.
[117] Waldek, S. (2022). NASA aiming for big 2023 thanks to generous budget request | Space. Space.com.
[118] Biden's 2023 defense budget adds billions for U.S. Space Force. SpaceNews.
[119] Office of Space Commerce (2022). FY23 Budget Proposes $87.7M for Office of Space Commerce.
[120] Marcia Smith. House appropriators are approving the big increase requested for the Office of Space Commerce. Twitter.
[121] Berger, E. (2022).We got a leaked look at NASA's future Moon missions—and likely delays. Ars Technica.
[122] Space Energy Initiative (2022). Space Energy Initiative.

Japan,[123] and India,[124] important U.S. strategies for the Artemis Program and In-Space Servicing Assembly and Manufacturing Strategies have yet to connect with Climate Change and Green Energy Goals, nor has the Administration sought to mobilize the latent energies of the space industrial base for Space Solar Power to achieve net-zero as it has for fusion.[125] Economic development and human settlement in space will clearly have the most profound, long-term impact on preserving Earth's biosphere. Choosing to remain terrestrially rooted with spiraling populations and industrialization is not a viable alternative. Moreover, China is on track to exceed the U.S. in science and technology investment by 2030 -- the U.S.'s share of global R&D spending declined 2% between 2010 and 2019, while China's share increased 7%. and U.S. federal investment in the science and technology needed to spark innovations crucial to the future of the U.S. economy and national security has fallen to 0.7% of our GDP today compared to 1.9% in 1964.[126]

**Bureaucracy** - The number of small satellite launches is expected to double from 428 in 2021 to 814 in 2022.[127] And yet, the federal agencies responsible for licensing and authorizing these satellite operations have not expanded or evolved sufficiently to address the demand (see Figure 25). In the case of the FAA's Office of Commercial Spaceflight, a planned budget increase was eliminated by House Appropriators at a time when significant expansion is essential to meet the need and demand.[128]

**Strategic Competition** -Insufficiently acknowledged in the west, China's strategic plan to displace the U.S. as the global economic power is forecast to occur by 2030 by some accounts.[129] China's economy grew by 7.8% in 2021 while the U.S. economy grew at 5.7%.[130] However, China's economy was recently projected to grow at only 2% in 2022 compared to the U.S. at 2.8%.[131] COVID lock-downs, an aging population and reductions in Chinese productivity leave room for the U.S. to out-pace China through pursuits of new markets with large growth opportunities. The new space economy is booming and is expected to exceed $1 trillion by 2040.[132]

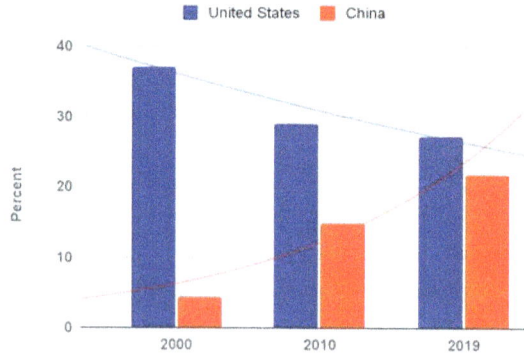

Figure 18. Technological competition between the U.S. and China in R&D is closing. (Credit: DIU)[133]

**Job Growth** - The demand for talent in the new space industry is off the charts right now and fueled by the tremendous amount of private investment pouring into the commercial space sector. Space Talent,

---

[123] Gislam, S. (2022). Japan to demonstrate space solar power by 2025 - Industry Europe. Industry Europe.
[124] National Space Society (2013). Space Solar Power: Key to a Livable Planet Earth. SpaceRef.
[125] White House (2022). Readout of the White House Summit on Developing a Bold Decadal Vision for Commercial Fusion Energy.
[126] Science and Technology Action Committee (2022). Our Nation's Leaders Can't Let China Win The Global Innovation Race. ScienceTechAction.
[127] Sriram, A. (2022). Capacity crunch may abort U.S. satellite boom as sanctions threaten Russia launches. Reuters.
[128] Foust, J. (2022). House appropriators reduce proposed budget for FAA commercial space office. SpaceNews.
[129] Rapp, N. & O'Keefe, B. (2022). China will soar past the U.S. to become the world's largest economy by 2030. Fortune.
[130] CEBR (2022). World Economic League Table 2022. CEBR.
[131] Bloomberg News (2022). Next China: US Could Finally Win GDP Growth Race. Bloomberg.
[132] U.S. Chamber of Commerce (2021). Space Economy: 4 Trends to Watch in 2022. U.S. Chamber of Commerce.
[133] NSF (2022). The State of U.S. Science and Engineering 2022. NSF.

an online talent network for the space industry, lists more than 25,000 jobs unfilled as of this writing, and the demand is expected to increase, despite flat or declining growth in other technology sectors.[134]

**Climate** - The U.S. is on track to reduce its total CO2 emissions up to 25% by 2030; however, that is only half of what is needed to comply with the goals of the Paris Climate Accord.[135] The U.S. must find other means to reduce up to 2.3 billion tons of carbon dioxide emissions by the end of the decade because 2.7 billion acres of new forests would have to be grown to remove that much CO2 from the atmosphere. The simple fact is that humans emit one hundred times more CO2 than all of the volcanoes on planet Earth combined. As the population grows, so does its net CO2 emissions.

*"We need to protect Earth and the way we will protect it is by going out into space."*
– JEFF BEZOS

**Critical Infrastructure** - The industrial base requires large scale infrastructure investment to develop here on Earth and in space. The U.S. does not have the critical infrastructure needed to support the new ecosystem. Whether it is a 'Range of the Future' (ROTF) able to accommodate a higher launch cadence with autonomous flight termination, or sufficient capacity and access to launch pads, processing facilities, test stands, or even more advanced capabilities such as hypersonic wind tunnels or closed cycle test facilities for nuclear rockets, the U.S. does not have the ground infrastructure required to support the growth U.S. industry is capable of. Moreover, the U.S. lacks the in-space infrastructure both below Geostationary orbit and beyond it to support growth, including propellant depots, Lunar landing pads, Cislunar navigational aids and space domain awareness sensors. The U.S. government should support the development of space infrastructure and should deem it critical infrastructure. Further, the USG should lead investment in space technology and space infrastructure in order to drive the future space and global economy and its growth.

*"Building transformative infrastructure is a matter of national greatness. With the right investments, America can retain its leading place in the world and remain a guiding force for the betterment of humanity.*  – ELI LEHRER, President, R Street Institute[136]

**Learning From History (the American Dream)** - The Lewis and Clark Expedition completed its exploration of the American West by 1806. It had multiple purposes -- to provide an economic survey, to expand scientific knowledge, and to thwart European nations from colonizing the western frontier of America[137] -- and opened a vast expanse to follow-on activity. Six decades later, the last spike connecting the Transcontinental Railroad connecting East with West was ceremoniously set as one of humanity's greatest achievements in the 19th Century. Six decades after Apollo, the United States is ready to drive the "Last Spike" enabling economic development and human settlement on the Moon (and beyond). President Abraham Lincoln justified the pursuit of the transcontinental railroad for national security, to move the Army into the new western territories expediently to defend them. Likewise, President Kennedy used the crisis with the Soviet Union over global influence to drive a North Star Vision and

---

[134] Jewett, R. (2022). How the Space Industry Competes Amid the Tight Market for Talent. ViaSatellite.
[135] Storrow, B. (2022). Hope Dims that the U.S. Can Meet 2030 Climate Goals. Scientific American.
[136] Lehrer, E. (2020). A Bold Vision for Infrastructure | National Affairs. National Affairs.
[137] McNamara, R. (2020). Why Did the Lewis and Clark Expedition Cross North America?. ThoughtCo.

employ all instruments of government, including the use of Other Transaction Authorities (OTAs)[138] as there was no other way to achieve the national security objectives on strategic timelines. A similar crisis for global influence exists today. Despite the near-term crises of their day, both leaders had powerful enduring visions for how these undertakings would influence future generations and reaffirm both American leadership and the American Dream. Neither the Transcontinental Railroad nor Apollo's mission of landing the first humans on the Moon had ever been accomplished before. Both pushed the limits of science and technology. Both inspired others to be like us. And yet, these two examples also illustrate the importance of scope. The Transcontinental Railroad was commercially led with significant government incentives and financing. As a result, it directly contributed to an accelerated growth and expansion of the U.S. economy beyond agriculture.[139] In contrast, the Apollo program was government-led and government dependent for sustainment. Both examples of grand strategy applied yielded extraordinary direct and derivative economic, development, prosperity, and security benefits that endured for generations.

> *"Human civilization is at a new moment of transition across social norms, economics, governance, and the environment, and is facing the dawn of a new era of interplanetary human migration (to Mars). In the future, historians will look to the first half of the 21st century to tell the story of how these changes started and unfolded through five domains of conflict."*
>
> –DR. LYDIA KOSTOPOULOS[140]

**Finance** - Private investment in commercial space companies grew by 48% over the previous year and expanded by 46% for a total of 241 transactions.[141] Initial Public Offerings via Special Purpose Acquisition Company (SPAC) transactions surged in 2021 but have since retracted considerably in 2022.[142] Unfortunately, there is very limited diversification in the types of financial investments available to the burgeoning new space economy. Debt financing and private equity investment are limited, but expanding slowly in this sector, contributing to the attractiveness of SPACs, when available.

## CAMPAIGN PLAN

**Space is the next great frontier for economic development and human settlement**. Like the exploration and settlement of the west, space offers new opportunities for realizing the American dream. The campaign plan to achieve the North Star vision for space must similarly resonate with all Americans. It will require advocacy, incentives and increased collaboration between the government and commercial industry. Equally important, it is imperative that the U.S. include allies and partners who share in our democratic ideals and vision for the future. Most importantly, we should diplomatically hold President Xi to his words that China upholds the principle of exploration and utilization of outer space for 'peaceful purposes' for the benefit of all humanity, and not simply as an extension of territorial claims and authoritarian rule.

---

[138] Defense Acquisition University (no date). OT Authority History | Adaptive Acquisition Framework.
[139] Hormats, R. (2003). Abraham Lincoln and the Global Economy. Harvard Business Review.
[140] Kostopoulos, L. (2022). The Emerging Domains of Conflict in the 21st Century. ORF Issue Brief No. 551.
[141] BryceTech (2022). Start-Up Space 2022: Update on Investment in Commercial Space Ventures. BryceTech.
[142] Nagle, M. (2022). Analysis: Days of Future SPAC—How SPACs Might Be ReWorked. Bloomberg.

> *"The commercial space industry is also a powerful engine for economic growth. Our nation's space economy employs over 354,000 people and generates $200 billion a year."*
> — VICE PRESIDENT KAMALA HARRIS, 2022[143]

A campaign plan aimed at achieving a North Star vision for space at an accelerated rate over an extended period of time should factor the following key objectives:

**Bi-Partisan Commission on America's Future in Space** - In order for a North Star vision to be enduring, it must transcend more than one administration. A bi-partisan commission established by the President and organized by the National Space Council could codify such a plan to ensure the U.S. retains its leadership, fosters alliances and ensures norms of behavior that are consistent with a democratic ethos.

**Prosperity Must Be Central** - The U.S. must retain its economic leadership and competitiveness in the 21st century. Therefore, the economic development of space resulting in the creation of new jobs and prosperity for a skilled workforce is paramount to the successful implementation of the North Star vision. This newly forged prosperity must extend to allies and friends of the U.S. as well. Much as the recent Indo-Pacific Economic Framework for Prosperity[144] provides a platform for mutual prosperity in an important terrestrial theater, we need an Economic Framework for Prosperity in Space, or a Cislunar Framework for Prosperity. Much as the U.S. Navy provides for freedom of commerce across the open oceans of Earth, the USSF may one day be necessary to assure the freedom of commerce across the vast Cislunar region and beyond.[145]

> *The bureaucracy must be incentivized and held accountable for fully implementing national strategies that derive from the Executive.* — SSIB'22 REPORT

**Streamline U.S. Bureaucracy** - Considering what is at stake geostrategically, it is unacceptable that U.S. companies have to wait many months, if not years, to receive words on the status of licenses, environmental assessments, or the operational authorities required to compete in the global marketplace. Complaints about bureaucratic delay or burden only seemed to intensify over the past year.[146] The bureaucracy must be incentivized and held accountable for fully implementing national strategies that derive from the Executive. Accountability requires clear lines of authority and clear delineation of responsibilities to implement the national space strategy, as well as, to promote U.S. economic growth and prosperity in the emerging commercial space sector. The Director of Space Commerce should be elevated as a position of strategic importance and report directly to the Secretary of Commerce rather than the National Oceanic and Atmospheric Administration (NOAA). The Federal Communications Commission (FCC) should have a finite amount of time to issue spectrum

---

[143] White House (2022). Remarks by Vice President Harris On Supporting the Commercial Space Sector. The White House.
[144] White House (2022). FACT SHEET: In Asia, President Biden and a Dozen Indo-Pacific Partners Launch the Indo-Pacific Economic Framework for Prosperity.
[145] Starling, C., Massa, M., Mulder, C., and Siegel. (2021). The Future of Security in Space: A Thirty-Year U.S. Strategy. Atlantic Council.
[146] Whittington, M. (2022). Why is FAA approval for SpaceX Starship orbital launches taking so long? The Hill.

licenses; the Federal Aviation Administration (FAA) should employ innovation to rapidly collect, process and address public comments; and the Environmental Protection Agency (EPA) should pursue an agile path of mitigation rather than restriction. In a word, the U.S. bureaucracy must be incentivized to find a way to "yes" while being held accountable for indecision which places jobs, companies and local economies at risk.

**Diversify Investments in Space** - As recommended in previous reports, financial engineering is needed to strengthen the viability of U.S. commercial space firms while mitigating risk posed by problematic foreign or non-standard investments. In addition, a center of excellence for the study of the new space economy would be helpful in developing the talent pool required to inform Wall Street and shape a sustainable future for long term investment in America's North Star vision.

---

*"Today, while the public may not fully realize it, the space ecosystem touches many aspects of daily life, abounds with commercial investment and increased commercial use cases, and plays a key role in advancing global sustainability and security priorities. It holds even greater potential for the future."* – McKINSEY & COMPANY[147]

---

**Diversify and grow the new space workforce** - The North Star Vision is 'not a space program, but a people program.'[148] As such, it is unattainable without a deep and diversified talent to achieve it. Direct investment in human capital is paramount to meet the needs of this and future generations. 'Human capital' refers to the economic value of a worker's experience and skills.[149] It includes assets like education, training, intelligence, skills, health, and other things employers value such as loyalty and punctuality. Winning the New Space Race requires a team-of-teams approach across a broad range of academic, industry and government sectors.

**Significantly mitigate Climate Change** - by leading the development of space-based solar power, moving heavy industry and mining off of Earth, and accessing strategic minerals key to the green economy, space development offers an additional powerful arrow in the quiver to combat climate change. Space development also enables concepts (such as the Sun-Earth-Lagrange 1 Sunshade) which could responsibly and reversibly conduct solar radiation management.

**Expand Artemis Accords beyond NASA** - Cislunar norms cannot be solely about civil space programs, but must encompass domain shaping for private and military actors. Agreements and institutions beyond the Artemis accords are required to play equivalent roles to the International Civil Aviation Organization (ICAO),[150] as well as a security community specific to space, as NATO is such a community specific for the transatlantic community.

---

[147] Brukardt et al (2022). The role of space in driving sustainability, security, and development on Earth. McKinsey & Co.
[148] See Statement for the Record of Charles E. Tandy, Prometheus Society, in support of H.R. 451 and 447 (1978).
[149] Kenton, W. (2022). Human Capital. Investopedia.
[150] Garretson, P. (2022). An ICAO for the Moon: It's time for an International Civil Lunar Organization. The Space Review.

## KEY ISSUES & CHALLENGES

**White House Vision Document Needed** - A White House-level document, equivalent to the Kennedy Administration's NSAM-144 (see Figure 13 at beginning of this section) that articulates America's broadest vision and approach for the settlement and economic development of space is needed to integrate and synchronize commercial, civil and national security space activities in a synergistic manner. No current document sets long-term goals for the whole-of nation or for the alliance and partners outside of China.[151] This undermines unity of purpose in the bureaucracies of the U.S. Government, who continue to argue regarding their larger mandate and focus. This may undermine the confidence of the electorate, which can see the deficit between the ambitions of its prominent citizens and the government. It undermines the confidence of industry that the government is 'on-side.' It undermines the clarity and ambitions which might attract international partners, including talented immigrants. Without clarity about contesting at the highest level of goals, many individual citizens wonder if they are playing for the B team.

> *"The fatalism of the limits-to-growth alternative is reasonable only if one ignores all the resources beyond our atmosphere, resources thousands of times greater than we could ever obtain from our beleaguered Earth."*
>
> - DR. GERARD K. O'NEILL, 1978[152]

**No Authorizing Legislation** - No current legislation provides the necessary authority, and clarity to enable and focus the responsible organizations on achieving the North Star Vision. For a brief period from 1988 to 1995, it was public law that, "Congress declares that the extension of human life beyond Earth's atmosphere, leading ultimately to the establishment of space settlements, will fulfill the purposes of advancing science, exploration, and development and will enhance the general welfare."[153] There have been more recent legislative attempts to enable NASA to "encourage and support the development of permanent space settlements," and to "expand permanent human presence beyond low-Earth orbit in a way that enables human settlement and a thriving space economy shall be an objective of U.S. aeronautical and space activities."[154] Nevertheless, there is a broad consensus objective that the long term goal of the human spaceflight and exploration program of the U.S. should be to expand permanent human presence beyond low-Earth orbit in a way that will enable human settlement and a thriving space economy, and that the best way to achieve this is through public-private partnerships and international collaboration, but it has yet to be codified in law.[155] This furthers the issues described above.[156]

---

[151] See China's Space Program: 2021 Year's Perspective (on the next page).
[152] Testimony of Dr. Gerard K. O'Neill, Princeton University, before the Committee on Science and Technology, U.S. House of Representatives, 25 Jan 1978.
[153] Congress (1988). Section 217.(a) of Public Law 100-685 100th Congress An Act To authorize appropriations to the National Aeronautics and Space Administration.
[154] Congress (2015). 114th Congress (2015-2016): Space Exploration, Development, and Settlement Act of 2016.
[155] NASA (2015). Pioneering Space National Summit 2015.
[156] Wolfe, S. & DeTorra, T. (2021). Op-ed | Space Settlement Act should guide Nelson's NASA tenure. SpaceNews.

Figure 19. Taikonauts for the Shenzhou-14 mission before launch to the China Space Station. (Credit: Licensed from Getty Images)

## 中国的太空计划：2021 年的视角
(China's Space Program: 2021 Year's Perspective)[157]

"To explore the vast cosmos, develop the space industry and build China into a space power is our eternal dream," stated President Xi Jinping. The space industry is a critical element of the overall national strategy, and China upholds the principle of exploration and utilization of outer space for peaceful purposes. Since 2016, China's space industry has made rapid and innovative progress, manifested by a steady improvement in space infrastructure, the completion and operation of the BeiDou Navigation Satellite System, the completion of the high-resolution earth observation system, steady improvement of the service ability of satellite communications and broadcasting, the conclusion of the last step of the three-step lunar exploration program ("orbit, land, and return"), the first stages in building the space station, and a smooth interplanetary voyage and landing beyond the Earth-Moon system by Tianwen-1, followed by the exploration of Mars. These achievements have attracted worldwide attention…

In the next five years, China will integrate space science, technology and applications while pursuing the new development philosophy, building a new development model and meeting the requirements for high-quality development. It will start a new journey towards a space power. The space industry will contribute more to China's growth as a whole, to global consensus and common effort with regard to outer space exploration and utilization, and to human progress. China will continue to improve the capacity and performance of its space transport system, and move faster to upgrade launch vehicles. It will further expand the launch vehicle family, send into space new-generation manned carrier rockets and high-thrust solid-fuel carrier rockets, and speed up the R&D of heavy-lift launch vehicles. It will continue to strengthen research into key technologies for reusable space transport systems, and conduct test flights accordingly. In response to the growing need for regular launches, China will develop new rocket engines, combined cycle propulsion, and upper stage technologies to improve its capacity to enter and return from space, and make space entry and exit more efficient. China will continue to improve its space infrastructure, and integrate remote-sensing, communications, navigation, and positioning satellite technologies. China will focus on new technology engineering and application, conduct in-orbit tests of new space materials, devices and techniques, and test new technologies in these areas: smart self-management of spacecraft; space mission extension vehicle; innovative space propulsion; in-orbit service and maintenance of spacecraft; and space debris cleaning…

China will continue to expand its space environment governance system… It will study plans for building a near-earth object defense system, and increase the capacity of near-earth object monitoring, cataloging, early warning, and response."

---

[157] PRC (2022). Select outtakes from China's Space Program: A 2021 Perspective. The State Council Information Office of the People's Republic of China.

## KEY INFLECTION POINTS

- **The U.S. fails to develop a North Star Vision or is Late-to-Need** resulting in less than full mobilization of the national energies, and loss of space superiority within a decade.

- **The U.S. develops a North Star Vision igniting industry and attracting new partners,** recapturing the initiative, and developing and securing a free and open Cislunar economy.

- **The Executive establishes a North Star Vision with the highest national priority** across the U.S. Government as President Kennedy did with Apollo in 1962.

*"Why the moon? Because the goal is Mars. What we can do on the moon is learn how to exist and survive in that hostile environment and only be three or four days away from Earth before we venture out and are months and months from Earth. That's the whole purpose: we go back to the moon, we learn how to live there, we create habitats."*

– SENATOR BILL NELSON, NASA Administrator[158]

## KEY ACTIONS & RECOMMENDATIONS

From the above observations and additional inputs from the working groups, participants advocated for or expressed interest in five overarching recommendations for action, which reflect the numerous inputs from workshop participants and their best assessment regarding which agency(cies) are in the best position to forward the necessary change:

### ATTENDEE RECOMMENDATIONS

**Publish a North Star Vision for Space Development** that is as clear and ambitious in scale and timeline as the PRC (see guidelines in last year's report). (OPR: National Space Council)

**Incorporate the North Star Vision objectives into an updated National Space Policy.** (OPR: National Space Council)

**Release implementing guidance to agencies to begin execution.** (OPR: National Space Council)

**Seek to legislatively codify the North Star Vision** in department and agency authorizations (NASA Authorization, National Defense Authorization, Department of Commerce Authorization). (OPRs: National Space Council, Domestic Policy Council)

**Open dedicated funding wedges to track development-enabling technologies** in the President's Budget for relevant agencies (NASA, DoD, DoC, DoT, DOE) to include In-Space Servicing, Assembly and Manufacturing, exploitation of space resources, infrastructure, and space mobility and logistics systems. (OPRs: National Space Council, OSTP, OMB)

---

[158] Smith, D. (2021). 'We're all citizens of planet Earth': former astronaut Bill Nelson on his mission at NASA. The Guardian.

Figure 20. LauncherOne released from Virgin Orbit's modified Boeing 747 aircraft named 'Cosmic Girl' demonstrating the flexibility of airborne launch. (Credit: Virgin Orbit)

# LAUNCH SERVICES

*"While the space economy is driven by the downstream revenues, the upstream segment, including launch services, is the backbone of the sector."*

– SPACETEC PARTNERS[159]

## BACKGROUND

Transportation and logistics are no less foundational to enterprise success in space as on Earth, be it in commercial, civil or national security domains. No nation can maintain a dominant position in the domains of air, land, maritime or space without a superior capability for rapid movement and sustainment that is both affordable and scalable. Historically, each introduction of a new transportation modality presents enormous economic opportunity, capitalized primarily by those first to market.[160] Before 1980, the U.S. launched the lion's share of global commercial payloads to space. But with the U.S. government as both the primary customer, and as landlord setting the operational rules at the launch sites, the commercial market sought more accommodating environments and they found them primarily at the European Space Agency launch site in South America, but also in Japan, Russia and elsewhere. By the beginning of this century, the U.S. market share in space launch services had all but disappeared. After a decade of innovation and significant private investment, the U.S. has begun to recover and is once again resurgent as a leader in space launch services. But competition remains - primarily from China and then Russia. The marketplace is ever changing and demands continuous innovation or history will repeat itself and the launch providers will seek greener pastures or simply fade away. That development would degrade the U.S. National Security posture. This New Space Race will be more contentious and more complex than the first. This new contest includes more than just two strategic competitors, involves a broad spectrum of powerful commercial interests, and with the field of engagement so expansive and pervasive, much more is at risk for America.

## CURRENT STATE

One of the key findings of the SSIB'21 Report, is that the space industrial base is assessed to be *tactically strong, but strategically fragile*.[161] This continues to be the case with respect to space launch services, which truly represents the cornerstone of the new space economy. No other area of the space economy receives as much private investment as space launch, with $23.3 billion in total investment since 2013.[162] As noted above, the new space economy is hinged to the evolution of disruptive new transportation modalities that transit to, from and within space. The U.S. has a healthy lead with the evolution of reusable launch vehicles and alternatives to traditional launch such as horizontal (airborne) and kinetic (spin) first-stage equivalents. That said, it is ill-advised to slow or stop the innovation that improves and

---

[159] SpaceTec Partners (2021). Space Launch Market Analysis. SpaceTec Partners.
[160] Rodrigue, J. (2020). The Geography of Transport Systems. Routledge. ISBN 978-0-367-36463-2.
[161] Olson, J. et al (2021). State of the Space Industrial Base 2021 Report. DIU, AFRL & USSF.
[162] Space Capital (2022). Space Investment Quarterly, Q2 2022. Space Capital.

disrupts the current means of accessing space because China continues to close the technological gap that once defined U.S. leadership in this sector. Working Group participants noted the following observations:

**Innovation is Reducing the Cost per Kilogram to Orbit** - The scope and diversity of launch vehicles (and alternatives to traditional methods of accessing space)[163] hold great potential to disrupt the launch services market. Multiple launch providers are developing and testing reusable launch vehicles while others leverage horizontal (airborne) or kinetic launch as alternatives to first stages. Collectively, this innovation is driving the cost per kilogram lower by a factor of five or more.[164] The resulting impact is closing business cases for new on-orbit services and growing the new space economy. This innovation is not just in the United States. Europe,[165] the UK[166] and Japan[167] all have announced projects for reusable rockets, and China appears to specifically be chasing a Starship class reusable launch vehicle of its own.[168]

Figure 21. Spinlaunch's A33 kinetic launcher at Spaceport America, NM. (Credit: Spinlaunch)

**Growing Need for Responsive Launch Capabilities** - As commercial, civil and national security users become more dependent on space systems and broadband communications architectures, the need for resiliency grows in importance. Responsive launch enables the rapid constitution of new satellites or spacecraft on orbit, as well as the timely reconstitution of satellites or commodities in mission-designed orbits to meet emergent needs.[169] DoD specifically requires both commercial and tactical responsive launch services on contract to support emergent requirements.

**Space Access is Increasing** - Year after year, the number of space launches from the U.S. (as well as other spacefaring nations) is rising at a non-linear rate. Meanwhile, the government capacity required to analyze and process licenses or assessments has remained relatively flat, contributing to bureaucratic delay inhibiting the speed of U.S. innovation in the face of strategic competition (See Figure 25 on page 43).

**Agility is Key to Competitiveness** - The evolution of commercial space launch technology continues to follow a spiral development cycle that seeks to incrementally improve successive vehicles. This practice is common across many high technology industries, but is relatively new within the space industry. As a result government regulators and licensors, organizationally and culturally aligned with legacy 'waterfall' engineering practices where change occurs over longer time cycles, have struggled to match the new pace of today's innovation.

---

[163] Wattles, J. (2022). SpinLaunch wants to radically redesign rocketry. Will its tech work?. CNN.
[164] Chow, D. (2022). To cheaply go: How falling launch costs fueled a thriving economy in orbit. NBC News.
[165] Joly, J. (2022). Move over, SpaceX: ArianeGroup to make Europe's first reusable and 'eco-friendly' rockets. Euronews.
[166] Pultarova, T. (2022). Orbex unveils reusable Prime rocket for small satellite launches from Scotland. Space.com.
[167] Nikkei staff writers (2021). Japan launches reusable rocket project, chasing Musk's SpaceX. Nikkei Asia.
[168] Jones, A. (2022). China could shift to fully reusable super heavy-launcher in wake of Starship. SpaceNews.
[169] Albon, C. (2022). US Space Force plan for rapid satellite launches may finally take off. C4ISRNET.

**Increased Demand for Payload Processing** - Workshop participants report that the availability of payload processing facilities (PPFs) is severely constrained. PPFs are environmentally controlled facilities where one or more spacecraft are readied and enclosed within a faring or deployment mechanism before integration with a space launch vehicle. This capacity can be addressed both by the government and industry.

**Limited Launch Complex Availability and Range Access** – The demand for range access from new and existing commercial launch providers may soon exceed the supply of available space launch complexes at both the Eastern and Western Ranges (Cape Canaveral and Vandenberg, respectively). Unless infrastructure improvements are implemented to increase capacity, this will create a bottleneck slowing U.S. space innovation.

**Growing Traffic Congestion in the Air and at Sea** - Launch and landing activities will increasingly affect other existing transportation networks. On the Eastern and Western ranges, the shutting down of the air and water space for every launch affects airline and mariners' routes, as well as, general aviation and recreational boating, whether the launch attempt is successful or not. These new activities will eventually impact other terrestrial transportation of roads & rail as the spaceport system evolves across the country. Real conflicts are arising with every launch activity at the Cape, and

Figure 22. First stage of Relativity's 3-D printed Terran 1 rocket at launch site. (Credit: Relativity)

they will inevitably occur elsewhere. Commercial airlines ask why they have to deconflict by rerouting or delaying flights with commercial launch providers; mariners do not keep track of commercial launch schedules; commercial launch providers do not notify mariners when they scrub their launch attempt for the day.

There is room for improvement. When the USG was the customer and there were few launches a year, these existing commercial enterprises would more readily defer to NASA's Exploration or DoD National Security Space missions. But they now ask why they should lose money so a commercial launch provider can make money?[170] These commercial aviation and maritime interests have a point, but equally important, they have a powerful voice in Washington and will be participating in the shaping of new policies.[171] Those voices will only become louder as the launch cadence continues to grow. A core function of governing is the management of conflicting industry priorities. The stakes of this competition to all sectors of the nation's transportation network demand this be done well.

---

[170] Zazulia, N. (2018). Industry: We Must Accommodate Commercial Space in the NAS — For Now. Aviation Today.
[171] Airlines for America (2019). A4A Joins Coalition in Letter to FAA Urging Safe and Efficient Integration of Commercial Space Operations into the National Airspace System. PRNewswire.

Technology is evolving and the key federal regulators (FAA and USCG) are working to improve notification and awareness, compliance and efficiency without risk to safety. However, the challenge is here and now. There needs to be smarter ways to manage air and sea space other than traditional NOTAMs[172] & NOTMARs.[173] The USCG is now utilizing QR codes on the NOTMARs for the Port Canaveral area to provide access to real time notification for mariners from the USSF (See Figure 23). That is particularly useful for the hundreds of recreational boaters and fishermen who ply the waters off the Cape and are not as attentive to restrictions as licensed professional captains. The FAA is working to better utilize technology and data analysis to limit restrictions of the National Air Space (NAS). A future with increased launch cadence drives the need to expand spaceports and look at alternative launch routes. A potential future goal is the establishment of clear corridors for spaceflight vehicles through the NAS where commercial and general aviation will be tasked to know if they are active or not.

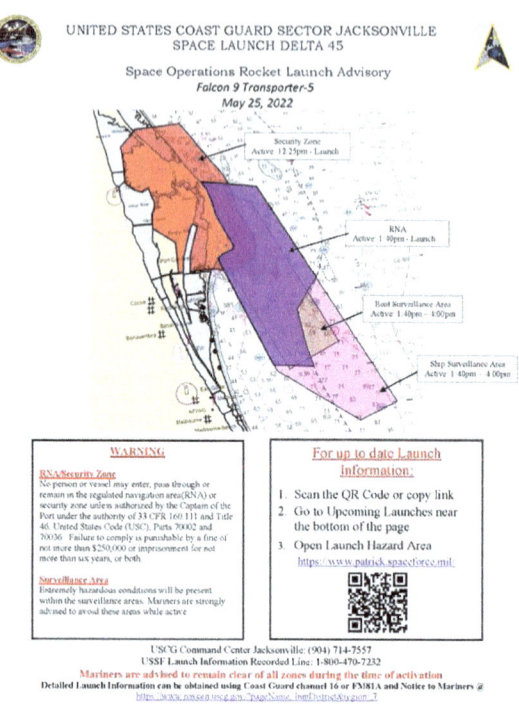

Figure 23. U.S. Coast Guard's improved Notice to Mariners (NOTMARs). (Credit: USCG)

## KEY ISSUES & CHALLENGES

**What's Driving Range of the Future** - The current capacity of U.S. space launch complexes is insufficient to meet the growing demand by U.S. commercial, civil and national security launch providers. If nothing changes, the U.S. will reach a choke point where the growth and responsiveness of launch operations will be artificially curtailed creating greater economic opportunity for strategic competitors (i.e., China and Russia). Launch, landing, and payload processing demands are up, stressing the availability of existing assets and facilities. Both Eastern and Western ranges, in particular, are seeing those demands increase rapidly, highlighting current infrastructure restrictions due to aging facilities and constrained readiness under existing operational procedures and policy. According to the USSF, the average age of range items with a potential mission impact and no available spares was more than 30 years old.[174] Several years ago, Cape Canaveral saw a couple launches a year; last year there were 31 launches of which 84% were commercial.[175]

---

[172] FAA (2022). FAA NOTAM Center.
[173] USCG (2022). USCG Navigation Center NOTMAR District 7.
[174] DoD IG (2022). Audit of DoD Maintenance of Space Launch Equipment and Facilities (DoDIG-2022-048). Department of Defense Office of Inspector General.
[175] Foust, J. (2022). Launch ranges lack spare parts to support growing demand. SpaceNews.

The U.S. launch infrastructure was built when the government was the only customer. Two decades ago, there began a partnership between the state and federal government entities to grow the capabilities at Cape Canaveral, FL.[176] That early attempt at collaboration continues to mature and has accelerated in the post Shuttle era.[177] There are several concepts being evaluated by different parties to address the same fundamental challenge. One option under consideration by the USSF PEO for Space Launch is to address funding shortfalls for maintenance and to support upgrades to newer capability. This would require new authorities from Congress to allow USSF to accept industry partner reimbursements to jointly fund infrastructure improvements[178] (a Public Enterprise Revolving Fund).[179]

Figure 24. Rocket Lab's twin LC-1A and LC-1B launch pads on the Mahia Peninsula, New Zealand. (Credit: Rocket Lab USA)

Space Florida, as a state partner, continues to evolve their planning process, implemented through the Florida Department of Transportation (FDOT) Work Program,[180] which annually solicits industry to project 5-year needs, currently identified as $2.8 billion,[181] in projected requirements to fund common use projects for basic utilities, road access, pads and processing facilities over and above estimated USG projects. The CSO, General Raymond, initiated a study in 2019 to assess different governance models for both the Eastern and Western range spaceports.[182] The stated intent was to free the USG and taxpayers from responsibilities associated with the support and nurturing of the growing commercial enterprises which would enable the USG to focus its resources and bandwidth upon its core missions. This issue will be addressed within the charter of the new National Spaceport Interagency Working Group under FAA leadership.[183]

Beyond new infrastructure funding, industry concerns have been expressed to address and clarify launch priorities within Range of the Future in the managing of competing requests for certain launch dates and range availability not only between USG and the commercial sector but increasingly between different commercial providers. Although these competing agendas are becoming prevalent at the Eastern and Western Ranges, the best practices being sought will find application at spaceports across the country as the industry develops.

---

[176] Florida Space Authority (2004). Briefing to the President's Commission on Moon Mars & Beyond. GovInfo.
[177] Florida Senate (2004). Florida Statute 331.367 Spaceport Management Council. Flsenate.
[178] U.S. Congress (2011). USC Title 10 § 2208. GovInfo.
[179] A revolving fund is a special account into which money is deposited for expenditure without regard to fiscal-year limitations. It must be specifically authorized by Congress.
[180] FDOT (2022). FDOT 5 Year Work Program.
[181] Space Florida (2022). Space Florida Board of Directors Agenda.
[182] Schoonmaker, R.L. (2020). A National Spaceport Strategy. Aerospace Corporation.
[183] FAA (2022). National Spaceport Interagency Working Group Charter.

**Failing Fast and Fast Learning** – Conflicting views on failure is inhibiting forward progress on launch safety. There is no greater challenge for the Eastern and Western ranges, and their government partners with landlord or regulatory responsibilities, than to transition away from the traditional mission assurance culture. That mentality had its purpose and was instrumental in aiding U.S. leadership in the early days of space transportation. But as history teaches us again and again, disruptive innovation is how leadership is sustained. The commercial sector is now driving U.S. leadership once again through rapid iteration and agility; where failure and learning from it are foundational elements of their culture for success. How will we ever get to the point that it is permissible to fly a booster over a populated area? Through the same way that the airline industry evolved to permit commercial passenger jets to fly over highly populated areas like San Diego, CA, through repeated demonstration of safe operations. We need to evolve commercial space launch in a similar fashion.

**Standards without Mandates** - There is an institutional tendency to establish standards that enable better utilization of facilities for launch and processing across a broad range of vehicles and payloads. Continuous innovation will one day enable the quick swap out of payloads on a ready vehicle at the last minute should one payload fail a final test. However, the industry is not yet at a point where firm vehicle and processing requirements can be imposed. Within what is a still nascent commercial market, enforcing dominant design criteria would stifle innovation.

**Barriers to entry for space startups are both huge and invisible** - The technical challenges of entry for startup and new entrants to the space industry are daunting enough without the impact and impediments of bureaucratic paperwork, be they from government landlords, regulators, or financial investors. These trials confront new companies at all spaceports, particularly those with federal landlords, and represent clear and present constraints upon the evolution of the U.S. leadership in space. This constraint is all the more stark when competing with strategic competitors who are not so shackled. Examples cited by Workshop participants include financial due diligence checklist requirements, obtaining Rough Order of Magnitude (ROM) estimates of costs, and the environmental permitting challenges[184] which arise without limit on time or costs before a possible decision is rendered. Then, there are challenges from the public, like the group suing the FAA and their decision to issue an operating license for Spaceport Camden. Camden County spent seven years and more than $11 million to develop the spaceport,[185] but the property has not yet been purchased or built. Now "the FAA [is] even more meticulous in their processing of all aspects of the application" (Landers), and companies say it's scary and daunting to face this risk.

**Insufficient Payload Processing Facilities at both USSF Ranges** - The number of bays required to support U.S. government space vehicle processing requirements alone is overmatched by current requirements driving the need for more commercial capacity by 2025.[186]

**Autonomous Flight Safety System is Critical to Range Transformation** - Autonomous Flight Safety Systems greatly enhance the pace of operations, and are proven to work. An example is the Autonomous Flight Safety System at work in New Zealand for Rocket Lab.[187] Of note, the New

---

[184] FAA (n.d.). Spaceport NEPA Requirements.
[185] Landers, M. (2021). Spaceport Camden clears one licensing hurdle. The Current.
[186] Presentation by USSF (SSC S3) at SSIB'22 Launch Services Working Group Meeting, 25 May 2022.
[187] Rocketlab (2019). Rocket Lab Debuts Fully Autonomous Flight Termination System.

Zealand spaceport is an FAA certified spaceport. Yet the U.S. has been unable to allow operation of such a system in the U.S., leading to a significant delay in flights from the Wallops Spaceport in Virginia.[188]

**Bureaucratic Delay is a Threat to Innovation** - For the U.S. to remain competitive in today's strategic environment, the U.S. bureaucracy must change to reflect the industry that it serves (see Figure 25), otherwise it risks killing the agile innovation culture that enabled the U.S. to become a leader in space technology.

**National Security Space Launch Contracts for All** - The procurement of space launch for national security systems should be expanded to include a broad range of space launch services and capabilities depending on an equally broad set of requirements. Diversity of launch options improves the resiliency of national security space systems.

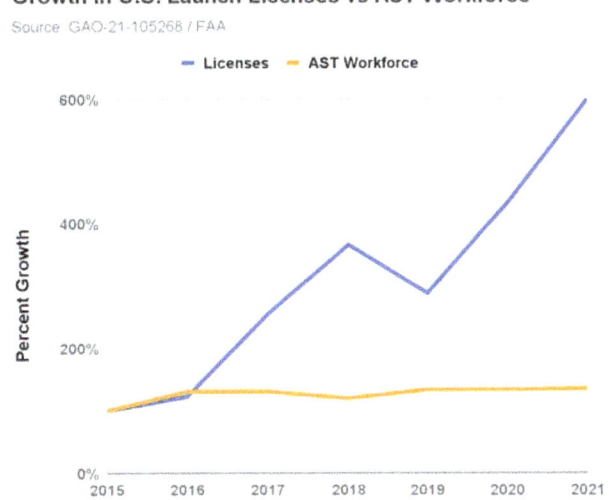

Figure 25. Growth in launch cadence unmatched by the FAA workforce that processes licenses. (Credit: DIU)

**Priority Licensing for National Security Enablers** - Commercial launch providers who are on contract to meet validated defense requirements, or contribute directly to national security priorities, should not be subject to bureaucratic delay. Developmental and operational activities should be allowed to continue with minimal constraint while companies work toward achieving compliance with non-safety related requirements or deliverables imposed by other government agencies.

## KEY INFLECTION POINTS

- **The first fully reusable launch vehicle flies successfully** from a U.S. spaceport demonstrating the viability and compatibility of performing developmental test and evaluation operations at commercial speed, and from someplace other than a high-density spaceport.

- **China exploits U.S. bureaucratic delay to close the gap in heavy lift, reusable launch capability.** Highly motivated and unconstrained, China will aggressively and systematically pursue disruptive technologies that undermine the U.S. innovation ecosystem.

- **The USSF successfully transitions its Eastern & Western Ranges into commercial spaceports** enabling an accelerated increase in launch cadence to support the growing new space economy.

- **Responsive launch services provide rapid constitution (or reconstitution) of space capabilities** essential to achieving resiliency in support of U.S. forces and their allies.

---

[188] Werner, D. (2022). State Fight: Virginia is for rockets. SpaceNews.

Figure 26. Starship stacked with a SuperHeavy Booster at Boca Chica, Texas, waits for a launch license to fly. (Source: SpaceX)

## Finding a way to "YES": Agile Policy enabling Agile Space

SpaceX thrilled the world with its rapid pace of test flights of its revolutionary, fully reusable Starship launch vehicle. But the last test flight of the Starship was performed on 5 May 2021. More than a year passed while the agile rocket company waited for a FAA Programmatic Environmental Assessment (PEA) to permit the initial two-stage flight tests from its developmental 'Starbase' in Boca Chica, TX. Starship's Boca Chica PEA had already been delayed five times, forcing the company to consider relocating its test flights to Cape Canaveral, the busiest spaceport in the U.S.[189] As of June, more than 75 compliance actions were required to fly.

In the meantime, China completed 55 space launches in 2021 with 60 more planned in 2022. China successfully launched its first space station; deployed many satellites; performed numerous test flights and technology demonstrations of remote sensing, communications and in-space logistics capabilities; and rapidly developed several new launch vehicles intended to compete with the U.S. commercial space industry.[190] More space launches occurred in 2021 than any other year in history and China contributed significantly to that achievement. Ironically, China is the worst polluter in the world by a factor of two while U.S. space launch CEOs are the most active in environmental sustainability.[191] Without a course correction, U.S. bureaucrats are unwittingly poised to 'win the battle, but lose the war' in this competition for America's space future.

The U.S. dominates the commercial space launch market today in large part because of the agile engineering practices employed by the most innovative new space companies in the world. However, agility requires a sustained ability to rapidly iterate the cycle of design, build and test activities where learning is facilitated through incremental success and sometimes failure. U.S. leadership in space will diminish under the burden of regulation and a bureaucracy that is ill-fit for today's strategic competition. Innovation and agility in policy making and regulating is required to meet the evolving demands of the 21st century. Appropriate waivers and exclusions should be exercised or considered, when necessary, in the best interest of U.S. economic growth and national security to remain competitive in this new frontier. Public safety, however, must not be compromised.

---

[189] Adalian, D. (2022). More Starbase delay. But SpaceX rockets ahead. EarthSky.
[190] Jones, A. (2022). China could shift to fully reusable super heavy-launcher in wake of Starship. SpaceNews.
[191] Bloomberg News (2021). The Chinese Companies Polluting the World More Than Entire Nations.

> *"As we bring more and more commercial providers on board, we're going to have more and more launches to get to that ultimate goal of two a day. It's all about converting our thinking and processes into an airport or services model in order to get there."*
> — BRIG GEN STEPHEN G. PURDY, Space Launch Delta 45 Commander[192]

## KEY ACTIONS & RECOMMENDATIONS

These recommendations reflect the numerous inputs from workshop participants and their best assessment regarding which agency(cies) are in the best position to forward the necessary change.

### SHORT-TERM PAYOFF

**Authorize a Public Enterprise Revolving Fund for Range Use Reimbursements** - There is no legal mechanism for the USSF (SSC/S3) to accept non-appropriated funds from commercial entities for the use and sustainment of DoD-owned and operated launch ranges. Approval of a fund is a necessary gap-filler until the government-owned western and eastern launch ranges transition to a commercial spaceport model. (OPRs: DAF, DoD)

**Automated Flight Safety System (AFSS) Approval** - Commercial launch providers continue to wait on the U.S. Government to approve Automated Flight Termination Systems (AFTS) solutions that are currently employed outside the Continental U.S. [including U.S. rockets at FAA approved ranges[193]] (OPRs: NASA, FAA).

**Airspace and Maritime Deconfliction** - Hazardous space activity, primarily launch and landing, must be safely and efficiently integrated into the airspace and sea lanes already thriving across the globe. Diverse commercial interests competing for the same place at the same time is a core government function. As the 'new kid', the commercial space business has found its seat at the table. But already there are major powerhouses in the airline and cruise and cargo industries who are making their voices heard in the policy setting landscape of Washington. Normalization of air, sea and space corridors for routine transportation operations is paramount to continued growth of the U.S. space economy. (OPRs: FAA, USSF, NASA)

Figure 27. Deconfliction of rocket launches from Cape Canaveral, FL, must be normalized with air and maritime traffic to sustain an integrated economy. (Credit: USAF)

---

[192] Cohen, R. (2021). Now boarding: Space Force wants to turn launch ranges into rocket 'airports'. Air Force Times.
[193] Rocket Lab (2019). Rocket Lab Debuts Fully Autonomous Flight Termination System.

**Quantity-Distance (QD) Standard Study** - Although there exists copious information on liquid natural gas (LNG)/methane primarily from USDOT[194] due to transportation of LNG via rail, pipeline, and truck, there are not the requisite studies to establish sound quantity/distance and Trinitrotoluene (TNT) equivalency on launch vehicles fueled with liquid oxygen (LOX) and LNG/Methane. Given the significant volumes of both contained in some of the new vehicles coming online very soon, having those standards in place is essential for the planning of spaceport operations. This cannot be an industry led effort as it is a core question of safety and a government responsibility. (OPRs: FAA, USSF, NASA)

**Agile Licensing & Advanced Analytics** - Government agencies responsible for issuing licenses or assessments need to innovate and adapt their internal policies and practices to accommodate an agile space industrial base. Agencies responsible for analyzing large amounts of data (including thousands of public comments) require modern software and data analytics tools to aid in processing licenses and resolving issues in an expedient manner (OPRs: NSpC, DoC, FAA, FCC).

Figure 28. Dream Chaser on final approach to landing following a resupply mission. (Credit: Sierra Space)

## MID-TERM PAYOFF

**Authorize and Fund the Range of the Future (ROTF) Initiative** - Spaceports provide our access to the tremendous resources available in the space domain. They will be the engines of economic opportunity[195] as well as sites where new challenges arise in the diverse arenas such as environmental impacts, deconfliction and encroachment. Range of the Future postures America to sustain its leadership in the new space economy and ultimately to enabling humanity to become a multi-planet species (OPRs: FAA, USSF, IAWG, NSpC)

**Acquisition Reform to Enable Commercial Service Contracts** - The USG is becoming ever more reliant on commercial services and capabilities, but the institutional acquisition process cannot yet find a way to: (1) effectively execute and fully benefit from meaningful market-based commercial service contracts,[196] (2) dramatically reduce proposals for services to 3-5 pages, and (3) develop a capacity for oral proposals to encourage immediate feedback for possible modification on the fly during

---

[194] U.S. Department of Transportation (2014). USDOT LNG Bunkering Update.
[195] Foust, J. (2019). Commercial spaceports increase focus on economic development. SpaceNews.
[196] Berger, E. (2017). Elon Musk knows what's ailing NASA—costly contracting. Ars Technica.

negotiations. This reluctance is an impediment to the USG fully benefiting from the very capabilities it most needs to meet its critical national security space requirements. It must establish clear needs, and then pay for them when provided. Industry needs requirements (i.e., don't specify what to build, but signal a desired outcome) and budgeting (e.g., budget for technology refresh in order to execute as soon as the technology is available) process to change and adapt to allow acquisitions to be more agile and faster. Not exploiting the full power of the greatest economy on the planet in the face of such menacing competition with China and Russia is to add risk, not reduce it. The Congress, White House, OMB and industry have been demanding an acquisitions process that is more agile and faster. (OPRs: DoD, NASA, FAA, Service Acquisition Executives)

## LONG-TERM PAYOFF

**Expand the pool of available commercial spaceports across the U.S.** - The economic benefits of space cannot be forever limited to coastal states. A safe and pragmatic plan is needed to chart a successful path enabling inland launch-to-orbit improving the scale and resiliency of space launch operations (OPRs: FAA, IAWG, NSpC).

**National Spaceport Strategy** - The new Interagency Spaceport Working Group seeks to map out a path to achieve operational integration of national (and ultimately global) spaceport systems using optimal governance models for federal ranges which leverage U.S. strengths. The charter is a good start,[197] but any possible progress will only be optimized through inclusion and participation by industry, states, and other stakeholders.[198] Transparency and clear goals must be assured for valid marketplace response, and visionary leadership from agencies is imperative for sustainable success in the New Space Race. (OPRs: Interagency Spaceport Working Group (ISWG), NSpC)

> *"US leadership was foundational in establishing the global air marketplace. As a result, every international air traffic control tower across the planet uses, of necessity, a common language; English[199]. In space there is no such common language for humanity to conduct business, yet. Space is hard enough, without having to do it in Mandarin."* - DALE KETCHAM, Space Florida, 2022

**Adopt a more optimal governance model for spaceports hosting federal missions** - whether for NASA Exploration or National Security Space, this has been explored for decades. Nowhere has that concept been more deliberated than the Eastern Range. The concept of a separate entity,[200] be it an independent authority,[201] a federal corporation[202] or some other construct which would free NASA and the military to focus limited resources on their core missions has been circulating in papers and studies since at least 2002[203] at the Cape. (OPRs: NSpC, DoC, FAA, NASA, USSF)

---

[197] FAA (2022). National Spaceport Interagency Working Group Charter.
[198] FAA (2021). Comstac National Spaceport Authority Report.
[199] International Civil Aviation Organization (2013). ICAO Language Proficiency Requirements. ICAO.
[200] NASA (2012). KSC Master Plan.
[201] Ketcham, D., Ball J. (2014). Is Space Exploration Best Served By Nasa Holding Property Assets As A Landlord?. 30th Space Symposium.
[202] Aerospace Corporation (2020). A National Spaceport Strategy (TOR-2020-00912). Aerospace.
[203] Interagency Working Group on the Future Management and Use of the U.S. Space Launch Bases and Ranges.(2002). The Future Management And Use Of U.S. Space Launch Bases And Ranges. NASA.

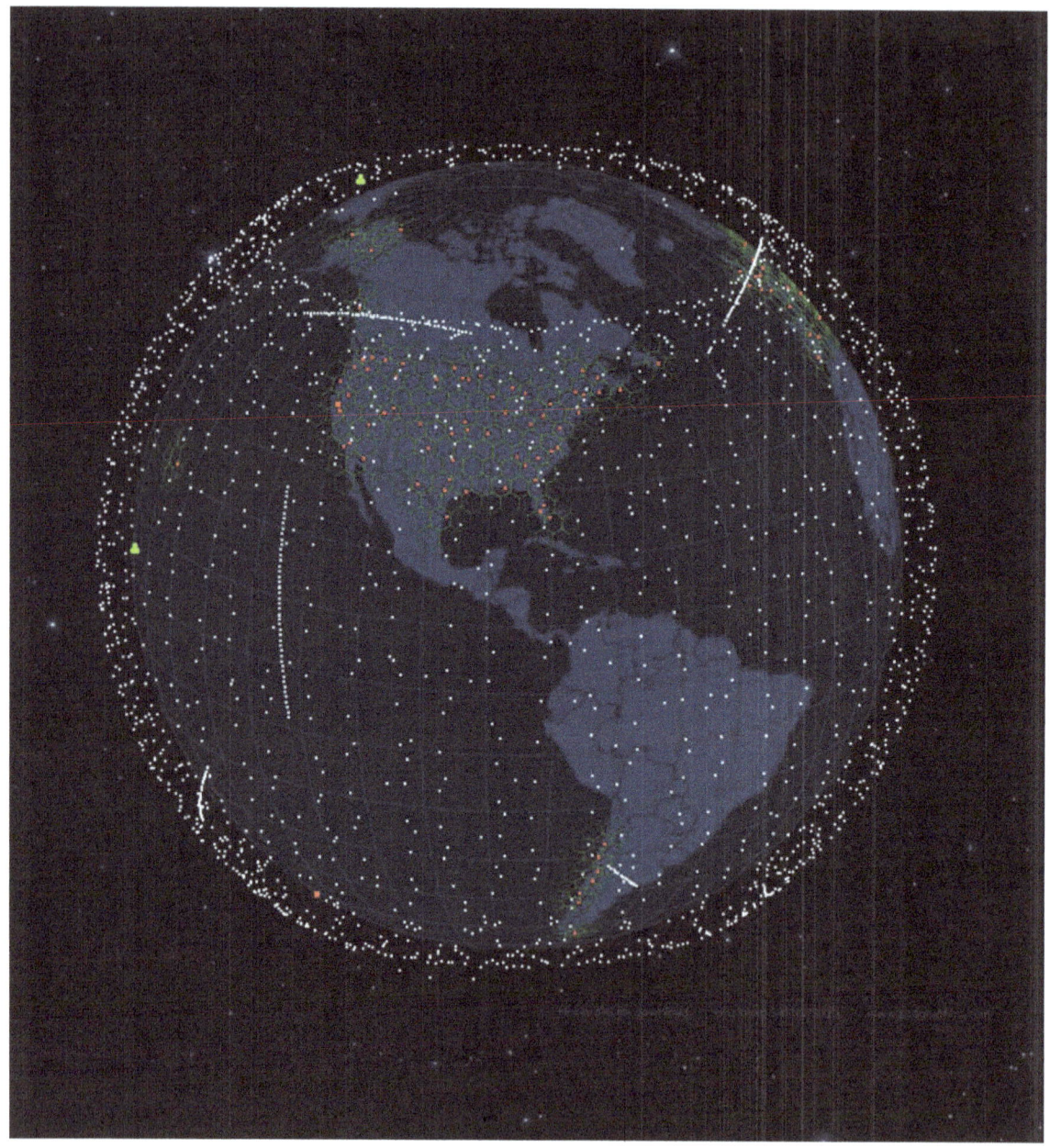

Figure 29. Real-time view of the Starlink broadband satellite constellation in LEO and terrestrial ground stations using the satellitemap.space web app. (Credit: Satellitemap.space)

# HYBRID SPACE COMMUNICATIONS

*"The committee is pleased that most national security space organizations have publicly embraced the Hybrid Space Architecture concept, notably the Space Force, National Reconnaissance Office, National Geospatial-Intelligence Agency, and the Space Development Agency. However, the committee further notes that funding for the Hybrid Space Architecture has historically lagged in budget submissions."*

– HOUSE ARMED SERVICES COMMITTEE, 2021[204]

## BACKGROUND

Since the State of the Space Industrial Base 2021, commercial space communications have truly proven their feasibility and utility to a hybrid Space Architecture (HSA). Almost all levels of leadership across the U.S. Government have acknowledged the need for proliferated hybrid communications for commercial, civil and national security applications. Significant commercial investment has already been made to offer secure, assured, multi-path voice/data communications 'as a service' across all of Earth orbits that is scalable to the Cislunar domain, and subsequently to the Earth-Moon-Mars system and beyond. How we shape this architecture as digital space infrastructure today will have a profound influence on its scalability, reliability and economic impact going into the future.[205]

The evolution of a Space Internet (some have proposed naming it 'the OuterNet'[206]) is seen as a natural extension of the Hybrid Space Architecture (HSA). Like the ARPANET, the intent of HSA is to facilitate a modern architecture that uses common protocols to facilitate data transport and processing in an Internet-of-Things environment to support joint all-domain command & control (JADC2)[208] and the future nuclear command, control & communications (NC3) architectures.[209] However, like the Internet, HSA is purposefully shaped to accommodate commercial, civil and national security space data transport leveraging open standards, zero trust and a revenue-positive business model that incentivizes industry toward common standards much like the terrestrial mobile telecommunications architecture. HSA's goal is "to integrate emergent commercial space sensor and communications capabilities with U.S. Government space systems while incorporating best-in-class commercial practices to secure and defend the network across multiple domains. The architecture will be secure, scalable,

*"TCP/IP doesn't work at interplanetary distances. So we designed a set of protocols that do."*

– VINT CERF, Chief Internet Evangelist for Google[207]

---

[204] HASC (2021). Report on H.R. 4350 NDAA, 10 Sep 2021. U.S. House of Representatives.
[205] B. Barritt and V. Cerf, (2018). Loon SDN: Applicability to NASA's next-generation space communications architecture. 2918 IEEE Aerospace Conference, pp. 1-9, doi: 10.1109/AERO.2018.8396643.
[206] Hitchens, T. (2022). Into the 'outerner': Secure 'internet in space' key to future Space Force hybrid architecture. Breaking Defense.
[207] D'Agostino, S. (2020). To Boldly Go Where No Internet Protocol Has Gone Before. Quartz.
[208] Chilton, K. (2021). The Backbone of JADC2: Satellite Communications for Information Age Warfare. Mitchell Institute.
[209] Cohen, R. (2019). Hyten: Future NC3 Network to Use Commercial Systems. Air Force Magazine.

responsive, and information centric. It must also be flexible to remain relevant and trusted during times of rapid technological change and dynamic threat environments. Software-defined controls, interfaces, and security are foundational to sustaining this agility. This architecture must be demonstrated as a payload (hosted or bespoke) capable of communicating across disparate government and commercial networks. To realize this demonstration, four domains have been identified: multi-path communications, variable trust protocol, multi-source data fusion, and cloud-based analytics."[210]

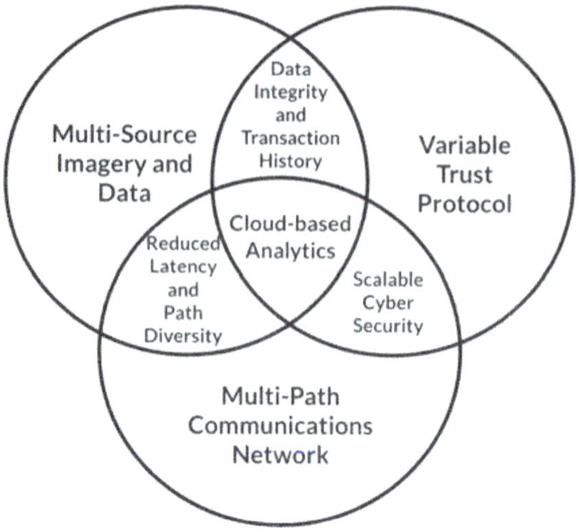

Figure 30. Key components of the Hybrid Space Architecture. (Credit: DIU)

## CURRENT STATE

**SpaceX Starlink** launched 989 new satellites in 2021[211] across 19 launches increasing their active subscriber base by 400%.[212] Starlink further announced partnerships with Microsoft Azure and Google Cloud Platform,[213] and proved itself rapidly deployable and surprisingly resilient to both jamming and cyber attacks throughout the Ukraine conflict.[214]

**Amazon Kuiper** has procured 83 launches for 3,236 satellites,[215] the largest such procurement in history, and plans to have half of its satellites deployed by 2026 according to its FCC license.[216]

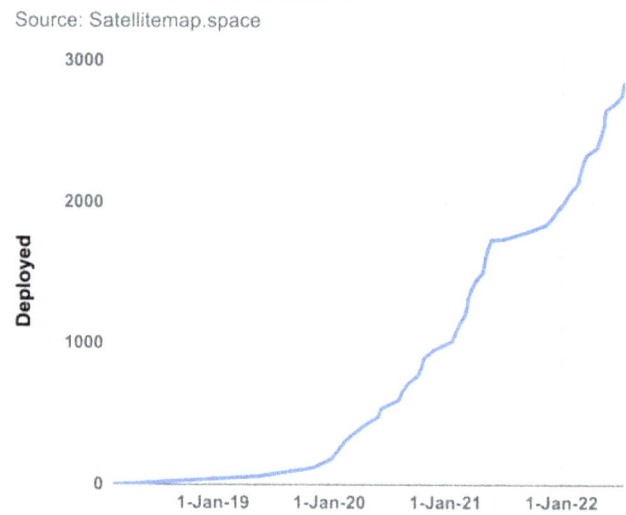

Figure 31. Starlink satellite constellation growth (Credit: DIU)[217]

---

[210] DIU (2021). Area of Interest: Hybrid Space Architecture (solicitation). DIU.
[211] SpaceX (2022). SpaceX - Launches.
[212] Kan, M. (2021). SpaceX's Starlink Now Serves Over 400,000 Subscribers. PCMag.
[213] Novet, J. (2021). Google wins cloud deal from Elon Musk's SpaceX for Starlink internet connectivity. CNBC.
[214] Jones, N. (2022). Thoughts on Ukraine as the Conflict Grinds On. National Defense.
[215] Amazon Staff (2022). Amazon makes historic launch investment to advance Project Kuiper. Amazon.
[216] Rivera, M. (2022). Project Kuiper Review 2022: Launch date, specs, and what it means for you. SatelliteInternet.com.
[217] McDowell, J. (2022). Starlink Statistics. Jonathan's Space Page.

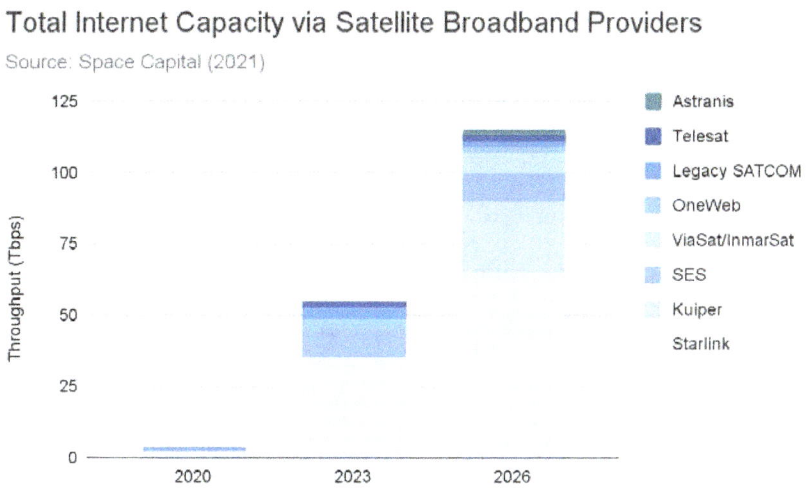

## Total Internet Capacity via Satellite Broadband Providers

Figure 32. Commercial SATCOM capacity aggregated by satellite broadband provider. (Credit: Space Capital)

**The National Reconnaissance Office** Commercial Systems Program Office (NRO/CSPO) signed 10-year contracts to procure commercial electro-optical imagery services from multiple providers.[218]

**NASA** selected six companies to participate in their Communication Services Program (CSP).

**The Ukrainian conflict** highlights the resilience and vulnerability of commercial communications - Early in the conflict, most Viasat ground terminals were rendered inoperable by Russian activities,[219] but within days, over 15,000 SpaceX Starlink terminals were shipped to Ukraine and rapidly deployed[220] reestablishing effective communications within the country despite Russian jamming and hacking attempts. Further, wearable 5G pucks with local mesh networking and short databurst satellite links drastically enhanced tactical situational awareness to field operators. It is evident to U.S. adversaries that commercial communications assets (in space and on the ground) are strategically and tactically important,[221] and thus going forward, we can expect measures will be employed by adversaries to disrupt these assets.

**Low-cost launch enables deployment** - The emergence of a robust U.S. launch supplier base, HSA-aligned satellite constellations (including SpaceX Starlink, Amazon Kuiper, and the SDA Transport Layer) can now be deployed economically.

---

[218] Datta, A. (2022). NRO Announces Billion-Dollar Historic Commercial Imagery Contracts to Maxar, Planet and BlackSky. Geospatial World.
[219] O'Neill, P. (2022). Russia hacked an American satellite company one hour before the Ukraine invasion. MIT Technology Review.
[220] Duffy, K. (2022). Elon Musk says SpaceX has sent 15000 Starlink internet kits to Ukraine over the past 3 months. Business Insider.
[221] Atlamazoglou, S. (2022). Ukraine says Elon Musk's Starlink has been 'very effective' in countering Russia, and China is paying close attention. Business Insider.

## KEY ISSUES & CHALLENGES

The key issues and challenges highlighted in the 2021 State of the Space Industrial Base report remain:

- **Acquisition and contracting** are evolving but remain poorly suited to purchasing data as a service.
- Acquiring **security clearances** is still an obstacle to greater participation by the industrial base.
- There is still **no central responsible office** for data acquisition leading to duplicated effort.
- Acquisition processes do not encompass all necessary **data types** yet.
- Lack of **exportability** of data is still hampering collaboration with allies and partners.

New challenges identified in the 2022 workshop include:

**Dual-use technologies** - how will the government, and especially the military, adapt to trusting commercially owned and commercially operated networks to transport sensitive data at varying levels of classification, potentially including weapons targeting information? Is industry willing to carry sensitive government data, potentially including weapons targeting information?

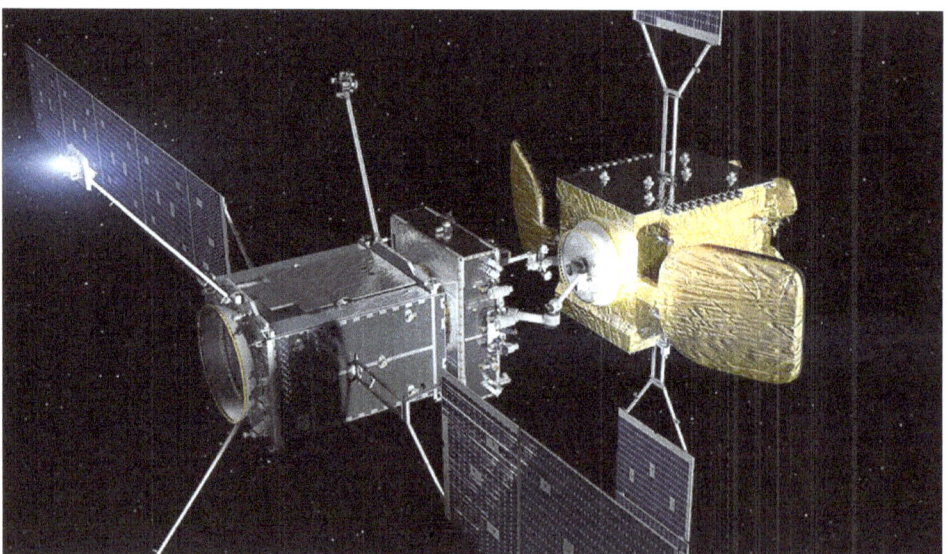

Figure 33. Autonomous rendezvous, proximity and servicing operations will rely on digital infrastructure for mission success. (Credit: Northrop Grumman)

**Supply chain** - the semiconductor shortage has demonstrated how fragile technology growth industries can become. New semiconductor foundries and associated primary resources need to be built on U.S. soil to ensure availability.

**Market oversaturation** - the economic downturn observed this year has demonstrated that venture capital may have been overenthusiastic in past years. The skepticism resulting from recent SPAC mergers suggests greater caution be used going forwards.[222]

---

[222] Foust, J. (2021). Space SPACs struggle to lift off. SpaceNews.

**Valley of Death Remains** - Despite significant growth in the government prototyping efforts, a valley of death still obstructs transition to programs of record. There are promising signs of government agencies moving towards service procurements with Operations & Maintenance (O&M) money as an alternative to traditional procurement, such as the NRO contracts described above and the Space Systems Command Commercial Satellite Communications Office (SSC/CSCO).[223, 224] However these programs are thus far each limited to acquisitions within a narrow domain (imagery for NRO/CSPO, communications for SSC/CSCO) without a comprehensive services program of record.

**Too many parallel efforts** - Perhaps due to Congressional emphasis on the Hybrid Space Architecture, there are now multiple parallel government efforts to establish such an architecture. These efforts may be duplicating work and confusing the industrial base.[225, 226, 227, 228, 229] A similar problem existed surrounding 5G adoption, so OUSD(R&E) established a centralized responsible Director of 5G Operations.

### HSA Statement of Principles

The Hybrid Space Architecture will dramatically improve deterrence and resilience in space while providing substantial new information advantage for science, commerce, and security.

The Hybrid Space Architecture will leverage:
- Multi-path, adaptative, secure communications; open mission systems; common standards
- Edge processing; autonomous command and control/tip and cue; artificial intelligence;
- distributed ledgers (e.g., blockchain)
- Terrestrial and space-based cloud infrastructure and analytics
- Commercial space manufacturing efficiencies (e.g., additive manufacturing and scale), systems,
- and data; digital modeling, design, and engineering; standards for cyber protection and secure
- supply chains; Agile/DevOps software and hardware approaches
- Low cost commercial bulk launch; responsive and resilient small launch, and
- New rapid government acquisition mechanisms to move quickly to the new architecture

This evolving resilient architecture will use a "variable trust" network framework for rapid and secure data exchange among proliferated satellite systems and services that are large and small; government and commercial; US and Allied; in various, diverse, and layered orbits. The architecture shifts from a platform-centric to an information-centric paradigm.

Table 1. Hybrid Space Architecture Statement of Principles (Credit: Smallsat Alliance)[230]

---

[223] Hitchens, T. (2021). Space Force Plans Up To $2.3B In COMSATCOM Contracts. Breaking Defense.
[224] Maucione, S. (2022). Space Systems Command using a 'buy first' attitude with procurement. Federal News Network.
[225] Erwin, S. (2021). seeks ideas for connecting government and commercial satellites. SpaceNews.
[226] Erwin, S. (2021). Viasat receives $50 million Air Force contract to develop space technology. SpaceNews.
[227] Space Development Agency (2022). SDA Seeks Industry Feedback through DRAFT Solicitation for NExT (National Defense Space Architecture (NDSA) Experimental Testbed).
[228] Hitchens, T. (2020). Space Force To Focus SATCOM Management On JADC2 Needs: EXCLUSIVE. Breaking Defense.
[229] DARPA (2022). Space-Based Adaptive Communications Node (Space-BACN).
[230] Nixon, S. (2022). Hybrid Space Architecture Statement of Principles. SmallSat Alliance.

## KEY INFLECTION POINTS

There has been some progress towards the key inflection points presented in the 2021 State of the Space Industrial Base:

**Establishment of a USSF commercial space services acquisition office** - Space Systems Command (SSC),[231] the NRO[232] and NGA[233] have each established their own commercial services offices.

**Adoption of hybrid space architecture standards** - SDA's optical communications standards have been widely adopted and successfully demonstrated by industry[234] while 3GPP have been successfully proliferating their terrestrial standards and rewriting them for the space environment.[235]

**Maturation of concepts of operations** - USSF stood up the Space Warfighting Analysis Center (SWAC) to undertake extensive force design of future space capabilities and concepts of operations.[236]

**Improvements in cloud analytics and edge processing** - The hardware is improving (supply chain problems notwithstanding), but this inflection point will inherently lag the transport layer.

**Long term leadership and funding** - As identified in the chapter featured quotation, there have been enthusiastic endorsements from leadership but lagging funding thus far. It was acknowledged that SWAC has been assigned some funding through a Program Decision Memorandum (PDM).[237]

New inflection points discussed in the 2022 workshop include:

**End-to-end demonstrations of HSA communications** and the technologies they enable to include JADC2 and Battle Management, Command, Control, and Communications (BMC3), real time tasking, collection, processing, exploitation, and dissemination tasking, collection, processing, exploitation and dissemination(TCPED) for sensor-to-shooter fires and beyond-line-of-sight tactical data links for situational awareness and shared targeting.

**Growing acceptance of Commercial Solutions for Classified (CSfC) cryptography** wherein the National Security Agency (NSA) accredits commercial solutions as trusted replacements for Type-1 cryptography.

**End-to-end demonstrations of data chain-of-custody** ensuring that data is complete and has not been tampered or observed.

---

[231] Hitchens, T. (2021). Space Force Eyes Buying Commercial Satellite ISR. Breaking Defense.
[232] SatNews (2021). NRO Director Reveals New Commercial Acquisition Opportunities Upcoming. SatNews.
[233] Underwood, K. (2021). NGA to Allow Commercial Sources as Primary. SIGNAL Magazine.
[234] Space Development Agency (2020). Space Development Agency Optical Intersatellite Link (OISL) Standard.
[235] 3GPP (2022). 3GPP Specification Set.
[236] Erwin, S. (2022). Space Force to take a fresh look at communications satellite needs. SpaceNews.
[237] Secretary of the Air Force Financial Management (2022). Department of Defense Fiscal Year (FY) 2023 Budget Estimates Justification Book Volume 1 of 1 Research, Development, Test & Evaluation, Space Force. SAF/FM.

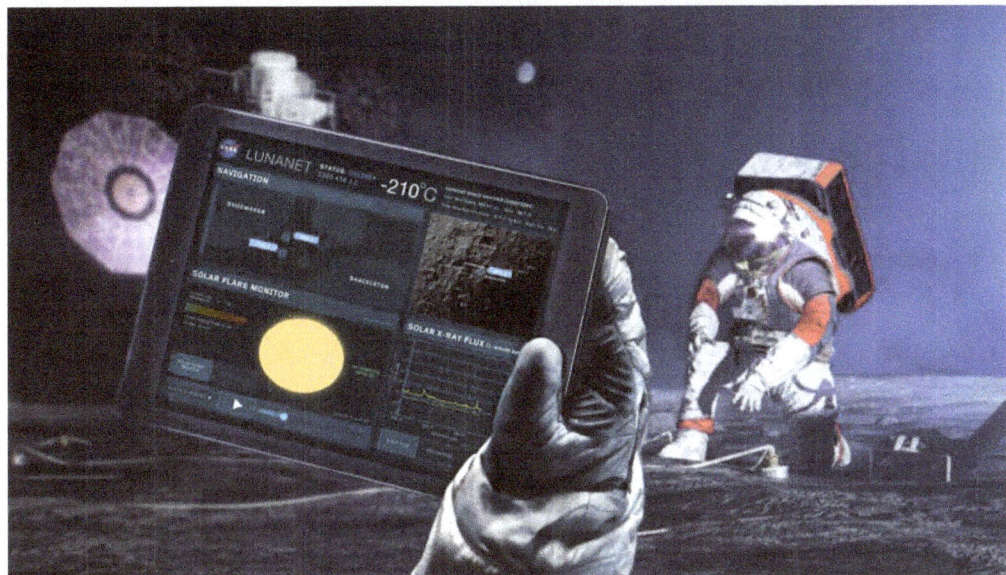

Figure 34. Artemis Astronauts will require wireless connectivity for safety and awareness. (Credit: NASA)

## The Case for the Outernet (Space Internet)

When astronauts return to the Moon for the first time in six decades, they won't be alone. Ideally, they will be connected to the Artemis Gateway, to mission control, to their rover, and to each other. They may even be connected via a software-defined network architecture that provides reliable, secure, low-latency communications, not unlike the internet on Earth but with one notable exception - the use of Delay/Disruption Tolerant Networking or DTN.

DTN is a set of standard protocols that make use of data streams and multi-path communications to successfully deliver information through a number of network nodes.[238] Data streaming is how information is commonly transported over the internet using TCP/IP. In space, however, there is no internet and TCP/IP is prone to errors when data streams are delayed or disrupted resulting in an error or loss of connectivity.[239] Two issues contribute to this problem in space: (1) the significant distance between nodes, and (2) the limited number of nodes constituting the network. Every device on the terrestrial internet is connected to each other by a very large number of wired and wireless connections permitting nearly instantaneous data relay between any two nodes. Errors that delay or disrupt data streams can be corrected quickly. Information traveling between Earth and Mars can take several minutes to transmit in one direction, so packet errors can simply time out. DTN addresses this by using a Bundle Protocol (BP) whereby each node can 'store-and-forward' streams without disruption if errors or disconnections occur. In this scenario, data streams literally 'hop' reliably from one node to the next.

DTN is beneficial not just for human missions like Artemis, but for robotic missions as well. DTN enables autonomous end-to-end data delivery based on headers within the data stream. In the absence of DTN, links between spacecraft have to be scheduled and configured well in advance of any data transmission. For this reason, DoD and NASA are both posturing to adopt DTN to enable the Hybrid Space Architecture, NASA's LunaNet and other software-defined networks that would benefit from the scalability and reliability of this technology. It's the first step in establishing the Outernet, a commercial internet in space, and creating new economic opportunities for all.

---

[238] Israel, D. et al (2019). The Benefits of Delay/Disruption Tolerant Networking (DTN) for Future NASA Science Missions. 70th International Astronautical Congress (IAC).

[239] Burleigh, S., Cerf, V. et al (2003). The interplanetary internet: A communications infrastructure for Mars exploration. Acta Astronautica, Volume 53, Issues 4–10. ISSN 0094-5765.

**The evolution from Radio Frequency (RF) broadcasts to optical multibeam** communications (to include terrestrial uplinks and downlinks) will avoid licensing and bandwidth restrictions and will reduce observability and the potential for interception and interference.

**HSA standards developed via industry consortia becoming widely adopted** and achieving critical mass such that they become self-perpetuating and trusted within industry, while fostering true interconnectivity between manufacturers. The widespread adoption into consumer focused industries including internet of things, autonomous cars, smart cities, cruise ships and airliners will drastically increase the technology pool, the talent pool, and the investment pool.

**Users rely on data from space-based sources** and data paths to develop new apps, generate new intelligence products, and work with data providers to add new sensor suites to the architecture without direct government involvement.

**New terminals bolster adoption of HSA** - Because the vast majority of devices tracked in the USSF Space Warfighting Analysis Center's (SWAC) force design are legacy user terminals and radios, reliant on narrow-band waveforms, that will be very difficult to integrate into the HSA.[240] A transitional period will be necessary with many "TranslatorSats" using hardware defined radios, cryptography and demodulators that packetize and route their data payload into the HSA. These TranslatorSats will then be phased out over years to come.

*"The Committee supports efforts to leverage commercial space networks to create an "outernet" for future military communications and believes the Space Force should undertake activities to promote interoperability standards and use of commercial ground and cloud architectures to increase the integration of commercial space networks."*

– HOUSE ARMED SERVICES COMMITTEE, 2022[241]

## KEY ACTIONS & RECOMMENDATIONS

These recommendations reflect the numerous inputs from workshop participants and their best assessment regarding which agency(cies) are in the best position to forward the necessary change.

### SHORT-TERM PAYOFF

**Programmatic Funding for Hybrid Space Architecture** - The size and scope of the Hybrid Space Architecture requires stable and predictable funding across the Future Years Defense Program. (OPR: Office of Management and Budget (OMB), USSF, NASA)

**Establish a forum to share Space Warfighting Analysis Center (SWAC) findings** via industry out-briefs at appropriate and accessible classification levels. Bolster and enhance those in the commercial arena at the TS/SCI level, but also implement the findings of the Classification Review of Programs of the Space Force[242] to ensure material is not over-classified. (OPR: USSF/SWAC)

---

[240] Hitchens, T. (2019). Is It Terminal? Mess Threatens SATCOM & Multi-Domain. Breaking Defense.
[241] U.S. Congress (2022). Report to Accompany H.R. 4432. House.gov; House Appropriations (2022). Appropriations Committee Releases Reports for Defense and Legislative Branch Bills, Fiscal Year 2023 Subcommittee Allocations.
[242] U.S. Congress (2022). Classification Review of Programs of the Space Force. FAS.

**Establish an independent industry consortium** to address OuterNet standards, incentivize interoperability, and monitor over-classification issues. The industry consortium should assign a representative to SSC/COMSO (Commercial Services Office)'s Program Integration Council (PIC).[243, 244] (OPR: Industry)

**Centralize OuterNet coordination within the DoD** to manage acquisition and prototyping activities across the DoD and to consolidate HSA related development (hardware/software, etc.), comparable to the OUSD(R&E) 5G Operations Office. Ensure the establishment of and coordination with the industrial consortium described above. Assign a representative to SAF/IA for international engagement. Coordinate with NSA and CYBERCOM to maintain consistent industry best practices across all networks, and regulate defensive cyber operations capabilities within orbiting assets and ground stations. Advise OSD on supply chain forecasting in coordination with the Space Information Sharing and Analysis Center.[245] (OPRs: USSF, OSD)

## MID-TERM PAYOFF

**Establish OuterNet as an explicit Program of Record** to ensure that HSA-related efforts are adequately prioritized and addressed and have a transition path. (OPR: HSA Coordination Office)

**Enhance the adoption of commercial innovations into operations** by standing up integrated operations/acquisition/test transition teams. (OPR: HSA Coordination Office)

## LONG-TERM PAYOFF

**Establish an OuterNet Working Capital Fund similar to the CSCO RF Broadband Working Capital Fund.**[246] Establish innovation programs to encourage adoption of new technologies into the HSA with ramped incentives for successful transitions. (OPR: Congress)

**Conduct roadshows to demonstrate various HSA-enabled capabilities and interoperability**, akin to existing industry "plug fests."[247] (OPR: OuterNet Industry Consortium)

---

[243] Space Force (2022). The Fight is on at Space Systems Command.
[244] Strout, N. (2021). Space Force will set up one office for commercial services, including SATCOM and satellite imagery. C4ISRNET.
[245] Space ISAC (2022). Space ISAC.
[246] Hitchens, T. (2020). Drafts New Acquisition Strategy For Commercial SATCOM. Breaking Defense.
[247] Bassler AG (2022). What kind of party is a Plug Fest | Software Development at Basler.

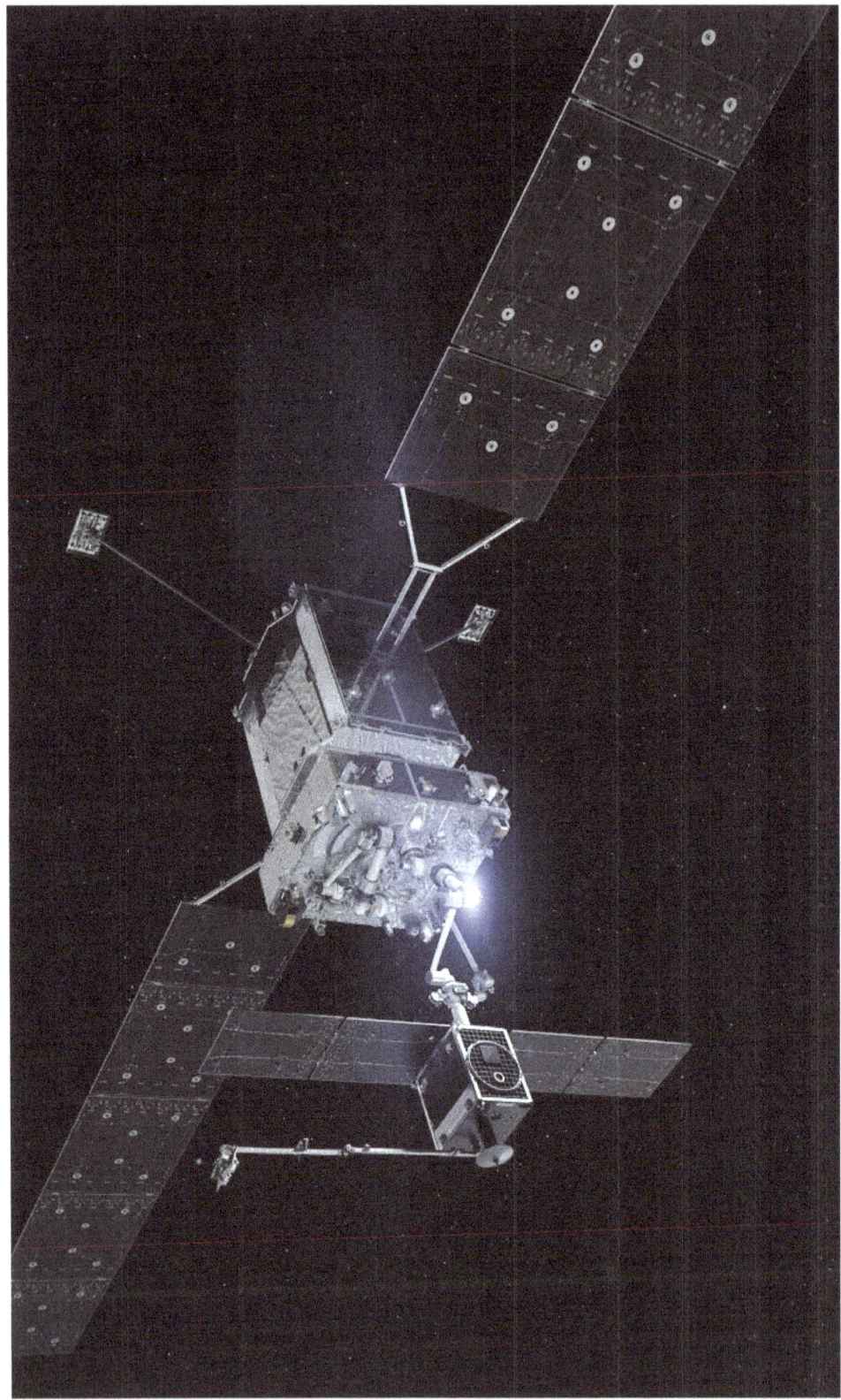

Figure 35. Space Logistics' Mission Robotic Vehicle (MRV) servicing a client satellite in geostationary orbit. (Credit: Northrop Grumman)

# IN-SPACE TRANSPORTATION & LOGISTICS

*"The United States recognizes the importance of establishing an ecosystem that promotes an affordable and sustainable (space) transportation network and logistics capability."*
— NATIONAL ISAM STRATEGY, 2022[248]

## BACKGROUND

**Space transportation and logistics capabilities are critical** for any expansive U.S. space future across civil, commercial, and military domains.[249] No nation has maintained a dominant position in a domain (air, land, maritime, space, and cyber) without a superior capability for movement and sustainment within that domain. The development of modular, serviceable, and reusable systems and sustainable in-space logistics infrastructure (both physical and digital) is central to expanding humanity's reach and the attainment of a robust space economy. To achieve leadership in this area, the U.S. must leverage the combined commercial, civil, and national security capabilities.

**Coming to Terms with In-Space Servicing Assembly and Manufacturing (ISAM)** - Many terms have been applied to this capability area, including NASA's On-Orbit Servicing Assembly and Manufacturing (OSAM), the broader Space Access, Mobility, and Logistics (SAML) initiative, the USSF's Space Mobility and Logistics (SML), and others, describing the broad array of technologies and missions required to support a robust and sustainable economy and presence in space.[250] Bolstered by an April 2022 release of the White House's ISAM National Strategy,[251] the wider space community is starting to coalesce around the umbrella term of ISAM.[252] Establishing a consistent terminology and understanding of the capability set involved is key to aligning strategy.

**A Space Logistics Chain** - Just as terrestrial logistics (land, sea, and air) serves as the backbone of the global economy, space will also require a logistics chain to support increased activity and to create a sustainable space ecosystem. While it is easy to dream up a future state with a widespread architecture of fuel depots and robotic space tugs to refuel spacecraft in diverse orbits, it is another thing to make this logistics chain a reality. The fundamental concepts and foundational technologies exist today, but there is more work and investment required to achieve a fully developed ISAM logistics chain.

**Interfaces and Modular Architectures Are Foundational** - The growth of ISAM will be dependent on the ability for services to support all users, commercial, civil, and national security. Common interfaces between systems and modular space architectures to accommodate interchangeability, upgrades, and ease of servicing are needed, and require definition and collaboration. Common interfaces and modular architectures are cross-cutting across the spectrum of ISAM capabilities and serve to support proliferation and growth in space, once they are fully and widely adopted.

---

[248] White House (2022). National In-space Servicing, Assembly & Manufacturing (ISAM) Strategy.
[249] Jehle, A. & Sowers, G. (2021). Orbital Sustainment and SML. Space Force Journal.
[250] Fernholz, T. (2022). US plans tech to refuel satellites and build factories in space. Quartz.
[251] White House (2022). In-Space Servicing, Assembly, And Manufacturing National Strategy.
[252] Rush, A. (2022). Op-ed | Assembling America's Future in Space. SpaceNews.

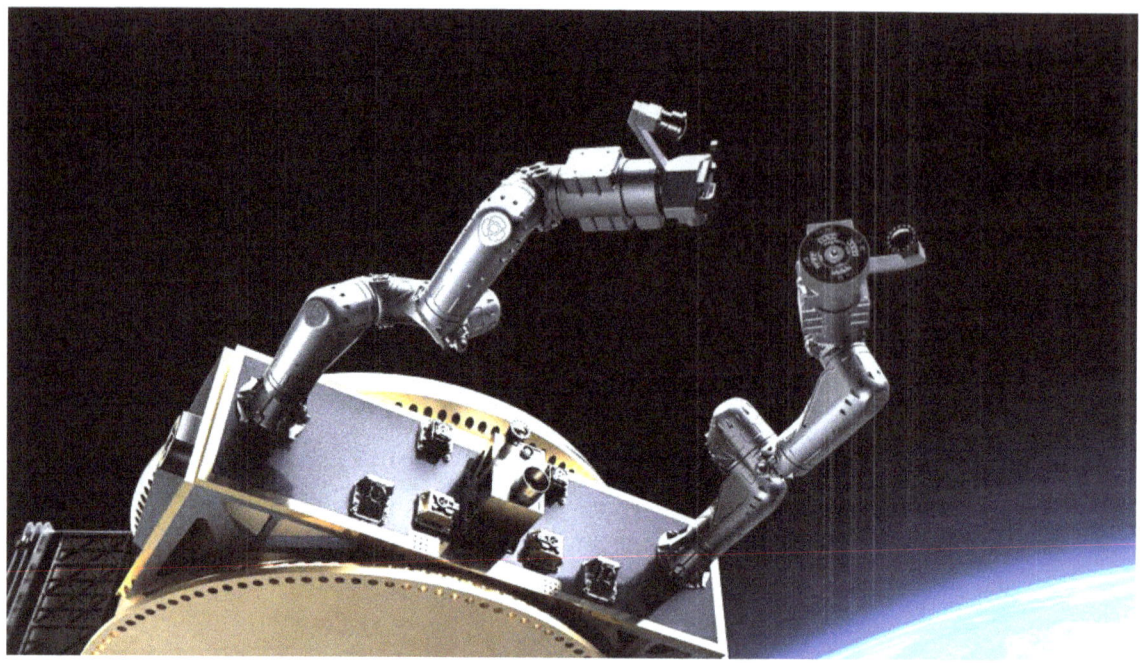

Figure 36. A variety of low cost, robotic manipulators are in development today to address the anticipated demand signal stemming from ISAM and logistics. (Credit: Motiv Space Systems)

## CURRENT STATE

**The U.S. Has Made Concrete Achievements** - The U.S. government, industry, and academia are all pushing on the ISAM front and there have been recent significant achievements. Building on a commercial demand signal for refueling, DIU released the Robust Access to Propellant In Diverse orbitS (RAPIDS) solicitation to establish the first bulk fuel depot and a commercially viable refueling service near GEO. This year, OrbitFab was awarded $6M from the U.S. Air Force and Space Force, and $6M from private investment through the AFWERX STRATFI program to integrate their Rapidly Attachable Fluid Transfer Interface (RAFTI) into DoD assets to allow them to be refueled.[253] Additionally, the SPACEWERX Orbital Prime effort is addressing orbital debris removal.[254]

Last year, Northrop Grumman successfully docked their Mission Extension Vehicle-2 (MEV-2) to the live commercial communications Intelsat 10-02 satellite to deliver life-extension services without interrupting service.[255] Advancements in digital enablers occurred this year, as Blue Origin solicited input from the community to create an open-source space robotics package called Space Robot Operating System (Space ROS), intended to simplify complete verification and validation and operate robots in space.[256] These achievements, and others, underscore that industry is starting to develop defense-relevant capabilities. Private capital infusion of small companies continued to grow, however, effort must be made to ensure these achievements and developments are made sustainable and transitional.

---

[253] Orbit Fab (2022). Orbit Fab announces $12M STRATFI Program — Orbit Fab | Gas Stations in Space™.
[254] Erwin, S. (2022). Space Force selects 125 industry proposals for on-orbit servicing technologies. SpaceNews.
[255] Northrop Grumman (2021). Northrop Grumman and Intelsat Make History with Docking of Second Mission Extension Vehicle to Extend Life of Satellite.
[256] Wessling, B. (2022). Open Robotics developing Space ROS with Blue Origin, NASA. The Robo Report.

**Continued Initiative and Drive for ISAM** - The White House released its ISAM National Strategy and NASA released the OSAM State of Play.[257] CONFERS continues to focus on interfaces and other servicing related capabilities,[258] and the Space Servicing, Manufacturing, Assembly, Robotics, and Transportation (Space SMART) Think Tank[259] brings together participants from industry, government, and academia to envision the future of space enabled by ISAM. Space Systems Command's Space Enterprise Consortium (SpEC) has released and funded prototype efforts with ISAM elements (e.g., Tetra-5,[260] CHPS[261]). However, these various strategy documents, working groups, and reports still have actions and implementation efforts to drive coordinated execution or establishing the "ISAM priority order" for development of capabilities.

**Growing list of Active Prototypes and Experiments** - Seed investments in new capabilities enabling modularity for in-space logistics are beginning to grow. The iBoss Intelligent Space Systems Interface (iSSI) was integrated and launched to the International Space Station this spring by SkyCorp for the purpose of maturing technology readiness and building flight heritage. The iSSI is a key component that has been developed commercially and adopted by a number of early stage space companies to increase interoperability. This also highlights the on-going need for commercial space platforms capable of supporting test and development in a variety of space environments.

Figure 37. Rendering of the Intelligent Space Systems Interface (iSSI) building flight heritage on the International Space Station. (Credit: SkyCorp)

**Foreign Powers Continue to Make Bold Moves** - Other countries are making strides towards developing ISAM capabilities, including ambitious programs such as the European Union's Prototype for an Ultra Large Structure Assembly Robot (PULSAR)[262] and the MOdular Spacecraft Assembly and Reconfiguration (MOSAR)[263] Horizon 2020[264] projects. China's Shijian-21 towed a dead satellite to another orbit.[265] Astroscale, a Japanese company, End-of-Life Services by Astroscale-demonstration (ELSA-d) successfully demonstrated its capability to repeatedly magnetically capture a client satellite.[266] The U.S. cannot assume that it will be the unchallenged leader in the future logistics chain.

---

[257] OSAM National Initiative (2021). On-orbit Servicing, Assembly, and Manufacturing (OSAM) State of Play. NASA.
[258] CONFERS (2022). About – CONFERS.
[259] Space SMART Think Tank (2021). Space SMART.
[260] Erwin, S. (2022). Space Force looking at what it will take to refuel satellites in orbit. SpaceNews.
[261] AFRL (2022). Cislunar Highway Patrol System (CHPS).
[262] European Commission (2021). Prototype for an Ultra Large Structure Assembly Robot. CORDIS; Pulsar (2022). Pulsar.
[263] European Commission (2021). MOdular Spacecraft Assembly and Reconfiguration | MOSAR Project | Fact Sheet. CORDIS.
[264] European Commission (2020). Horizon 2020.
[265] Jones, A. (2022). China's Shijian-21 towed dead satellite to a high graveyard orbit. SpaceNews.
[266] Astroscale (2021). Astroscale's ELSA-d Successfully Demonstrates Repeated Magnetic Capture.

## KEY ISSUES & CHALLENGES

Although progress has been made in fielding new capabilities in support of in-space transportation and logistics, challenges remain in leveraging these capabilities to create a sustainable ecosystem. Mission concepts and spacecraft must be architected differently, and with careful consideration of how to encourage and foster commercial development.

**Standards Continue to be a Risk** - Standards can be enabling for the future space logistics chain, allowing for interoperability, alignment across ISAM verticals, and supporting competition. The challenge is that currently there is insufficient consensus on standards, either from industry or the U.S. government. The result is a situation where lots of interfaces exist, but no clear driving standard enables universal interoperability. In other cases, lack of standards have slowed progress and development of key interfaces. For commercial industry, this presents a business risk given the uncertainty of if or when the U.S. government may adopt and push a standard interface. From the U.S. government perspective, the various non-interoperable interfaces developed across industry, academia, and government presents an "analysis paralysis" problem, waiting to see what interface may be the winner before adopting it into future architectures.

---

*"Economic growth requires innovation. Trouble is, Washington is practically designed to resist it. Built into the DNA of the most important agencies created to protect innovation, is an almost irresistible urge to protect the most powerful instead."*

– LAWRENCE LESSIG, Harvard Law School, 2008[267]

---

**The Space Logistics Chain is Farther Away Than Anticipated** - Pieces of the space logistics chain are being developed and demonstrated, but there is no coordinated effort to bring them all together. To achieve the full space logistics chain, everything from increased upmass to orbit, fuel depots, in-space transfer vehicles, robotic assembly, and space domain awareness (among others) is needed. Without an end state identified, and coupled with strategically determined investment and development, the various prototypes and technology demonstration missions for these capabilities risk being "on-offs" and abandoned, before supporting growth of the logistics chain.

---

[267] Lessig, L. (2008). It's Time to Demolish the FCC. Newsweek.

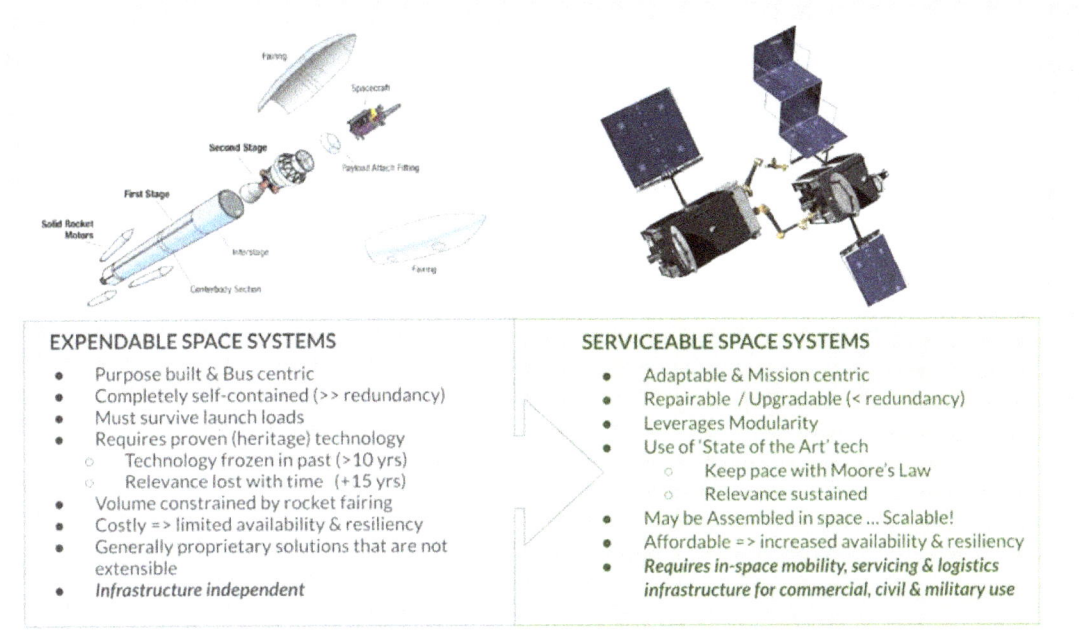

Figure 38. In-space Servicing, Assembly and Manufacturing (ISAM) and in-space logistics are critical enablers in the transformation of space operations to modular, sustainable and scalable systems. (Credit: DIU)[268]

## Transforming Space Operations with ISAM and Logistics

Imagine flying a commercial airliner to Europe and then throwing it away at the end of the journey. This is a simple but effective analogy to describe the challenges in the early days of space operations due to the tyranny of the rocket equation which mathematically tells us that generally no more than 4% of an expendable rocket's total mass is usable for its intended payload.[269] The remaining mass is primarily fuel, engines and structure. The cost of producing and flying large expendable rockets is prohibitive to achieving frequent access to space, or flying anything more than the absolutely minimum mass to accomplish a mission. As a result, anything worth sending to space had to be self-contained and highly resilient. With the advent of heavy lift, reusable launch vehicles, this paradigm is about to be disrupted for good.[270] Reusability significantly reduces cost and increases cadence of flight operations enabling new operational concepts, and for many business cases to close with revenue positive outcomes. With routine access, future satellites can be serviced and upgraded to retain their technological relevance. Rather than creating debris, parts or entire systems can simply be moved, recycled or returned to Earth, the Moon or elsewhere. Future satellites will be more like cell towers that serve modular components that are attached and replaced based on evolving needs.

The amazing images from NASA's James Webb Space Telescope (JWST) should convince us of the necessity to master this ability to assemble, service and sustain large space systems for the future. JWST had to be built to deploy from an expendable rocket requiring it to be folded like an origami bird within the rocket's fairing. It had to be designed to survive the g-forces during the first few minutes of launch, and include redundant components ensuring reliability and success over the lifespan of the telescope. In other words, JWST had to be an exquisite system - a one of a kind telescope. ISAM will enable future space telescopes and other spacecraft to be built and sustained more affordably and in greater numbers. In-space logistics capabilities will provide the means to deliver parts, commodities and the robotics (or astronauts) to render repairs or upgrades. Such systems can be orders of magnitude larger, more capable and more resilient. Space operations based on ISAM plus logistics will facilitate even greater achievements for all humankind.

---

[268] Richards, M. (2006). On-Orbit Serviceability of Space System Architectures. MIT (Masters Thesis).
[269] Zur, C. (2020). Escaping the Tyranny of the Rocket Equation. Scientific American.
[270] Simberg, R. (2021). Walmart, But for Space. The New Atlantis.

**Spacecraft Today Are Not Designed for ISAM** - No mandates currently exist with strict enough enforcement to ensure spacecraft are designed for future servicing, missing an opportunity to provide a foundation for development and innovation. At best, Congress mandated that future space astronomy missions are serviceable, but even this was qualified with a statement of "if practicable and appropriate" allowing it to be circumvented.[271]

**Lack of Funding Support** - Budgets and investment can speak louder than words. Of the $756B DoD budget in FY2022, $18B was allocated for the USSF, with only a small fraction of that for ISAM related efforts.[272] In the private sector, investments in ISAM-related companies accounted for only approximately $300M in 2021, dwarfed by the almost $53B injected into satellites and launch companies.[273] If ISAM capabilities are needed and desired, the signals required to promote innovation and funding are not yet there.

## KEY INFLECTION POINTS

**Wide adoption of common interfaces and modularity** to encourage growth of interoperability, competition and open systems architectures.

**Government establishes itself as an anchor tenant but not the only tenant**. Providing a clear and consistent demand signal from the government will allow commercial industry and private investment to build upon it without being solely reliant on government funding.

**Government selects a flagship mission to require ISAM capabilities** to help kickstart the ecosystem development across multiple capability areas. Not just another demo, this mission would allow the more nuanced ISAM benefits, such as modularity and staggered launches for phased missions allowing reduction of the missions "standing army," to be tested out and create a new format for an ISAM Program Life Cycle.

**Development of breakthrough advanced propulsion capabilities** to open the floodgates for larger, more maneuverable platforms.

**Government mandate for incorporation of modularity and servicing into spacecraft design**. If all government spacecraft moving forward have servicing in mind, this creates a quantifiable demand signal for ISAM services and development.

**Support for on-orbit demos** that are allowed to take risk and fail will elevate the Technology Readiness Level (TRL) of ISAM technologies and lower concerns to enable faster innovation and adoption.

**Strategic competitor achieves logistics chain dominance first**. If the U.S. does not lead in ISAM and other countries keep advancing, the U.S. might be beholden to the ways of practice, standards, and interfaces adopted by the larger space community first, stifling innovation and growth of the U.S. industrial base.

---

[271] Thronson, H., B. Peterson, M. Greenhouse, H. MacEwen, R. Mukherjee, R. Polidan, B. Reed, N. Siegler, H. Smith (2017). Human space flight and future major space astrophysics missions: servicing and assembly. NASA NTRS.
[272] Erwin, S. (2022). Biden's 2023 defense budget adds billions for US Space Force. SpaceNews.
[273] Space Capital (2021). Space Investment Dashboard. Space Capital.

# KEY ACTIONS & RECOMMENDATIONS

These recommendations reflect the numerous inputs from workshop participants and their best assessment regarding which agency(cies) are in the best position to forward the necessary change.

## SHORT-TERM PAYOFF

**Government should be advocating for common interfaces, not forcing standard interfaces.** Work with consortiums and similar organizations to establish standards for interfaces but allow for industry to lead in actual interface development and innovation. (OPRs: CONFERS, NASA, USSF, DIU)

**Improve coordination on ISAM strategy** to establish the capability priority order and avoid duplicative efforts. (OPRs: NASA, USSF, OSTP)

**Reduce classification levels for commercially developed capabilities.** Industry is working to commercially achieve defense relevant capabilities, which presents an opportunity to lower the barriers and promote innovation and collaboration with DoD. (OPRs: USSF, NRO, DIU, AFRL, DARPA)

## MID-TERM PAYOFF

**Government mandates that spacecraft are to be designed for ISAM** to establish a baseline of demand for commercial industry to develop around. (OPRs: NASA, USSF, AFRL, SDA, NRO)

**NASA and DoD improve the transition path from demonstrations and prototypes to operational capability.** Each demo should be viewed as a new tool in the ISAM toolbox and fully utilized to test out further ISAM capabilities and bring down the risk of new operations. (OPRs: NASA, USSF, DIU, AFRL, DARPA)

## LONG-TERM PAYOFF

**Government improves its demand signal for private industry** to build upon through consistent and continued funding and program commitments, and demonstrating that if ISAM services are created, the government will be a customer. (OPRs: NASA, OSTP, USSF, DIU)

**Government shifts focus towards a logistics chain end game** versus individual capability demonstration. By setting its sights on achieving technical maturity on the right mix and priorities of ISAM capabilities, the government can set out the pathway towards a fully realized space logistics chain. (OPRs: NASA, USSF, DIU, DARPA)

**The U.S. leads the modernization of space laws and treaties**, most of which did not contemplate commercial activity in space. The U.S. can either seek to reform existing frameworks or forge new multilateral agreements for rules of engagement, such as the Artemis Accords. (OPRs: OSTP, State Department, Congress)

Figure 39. Nuclear propulsion will make the solar system more accessible. (Credit: USNC Tech)

# NEXT GENERATION POWER & PROPULSION

*"We keep starting and stopping ... We may have spent as much as $20 billion on developing (but never flying except once) space fission systems."*

— BHAVYA LAL, NASA[274]

## BACKGROUND

In the last year, the U.S. Government has made progress towards a National Strategy for Nuclear Power and Propulsion in Space.[275] On the civilian side, both the human missions themselves and the logistics missions that support them will require large quantities of propellant and highly efficient propulsion systems to achieve affordable and routine access to the Earth-Moon-Mars system. Propulsion and power technologies such as next-generation radioisotope, fission, and fusion power systems, high-power electric propulsion, nuclear thermal and nuclear electric propulsion, and Lunar-sourced propellant must all be developed to support new DoD mission sets in Cislunar space. These new technologies will also sustain the future spaceflight ecosystem and enable civil human spaceflight and exploration goals. This report summarizes progress made over the last year and challenges to be addressed to expand mobility and energy supplies more rapidly, including regulatory and licensing obstacles and a lack of ground testing facilities. Nuclear and solar power and propulsion, power beaming and alternative propellants will be among the innovative technologies explored and discussed.

**Rocket Propulsion and Delta-V** - Advanced propulsion technology is essential to expanding the capabilities of spacecraft. The ability for a spacecraft to change its velocity (called Delta-V), is directly related to the propellant choice. Current state-of-the-art in-space propulsion systems based on chemical propellants may not meet requirements of 21st century NASA and DoD space missions. While the proliferation of refueling stations in Earth orbit may help legacy chemical fuels stay relevant for the next few decades, as military and civil missions expand into a nearly 2,000x larger Cislunar space (vs GEO volume),[277] it is unlikely that fuel depots will always be available and in the right spot. More efficient

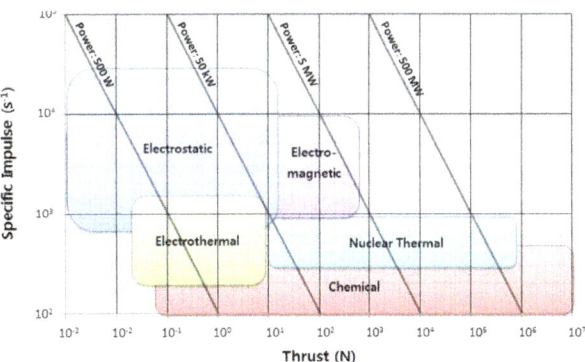

Figure 40. Specific Impulse and Thrust operating regimes. (Credit: Ebersohn et al)[276]

---

[274] Lal, B. (2022). Keynote Presentation: Going Nuclear - Prospects for Power and Propulsion in Space. SSIB'22 Workshop.
[275] White House (2020). National Strategy for Space Nuclear Power and Propulsion. Office of Space Commerce.
[276] Ebersohn, F. et al. (2013). Resistive Magnetohydrodynamic Study of Magnetic Field Effects on Plasma Plumes. 44th AIAA Plasmadynamics and Lasers Conference. DOI: 10.2514/6.2013-2759.
[277] Holzinger, M. J., Chow, C. C. & P. Garretson (2021). A Primer on Cislunar Space. AFRL.

and powerful propulsion systems are needed to ensure maneuverability in this domain, and to maintain a mobility advantage.

**Powering Spacecraft Systems** - Similarly, continuous and long duration electrical power generation sources are needed to fulfill future DoD mission sets in Space. These portable power sources are required in applications where other power sources such as solar cells and chemical batteries are not suitable. As the DoD Space domain expands to Cislunar areas,[278] and new missions demand more electrical power, it's clear that lightweight and long duration nuclear power sources are a key enabling technology. Current devices suffer from fundamental limits in power conversion efficiency. For example, a solar-powered spacecraft with an active radio frequency (RF) payload (e.g., synthetic aperture radar) may only be able to operate 20% of the time due to limitations on solar irradiance. Next generation power technologies are needed to enable spacecraft to run even more demanding payloads (e.g., directed energy) at higher duty factor to maximize both military and commercial utility.

Figure 41. DARPA's Demonstration Rocket for Agile Cislunar Operations (DRACO) seeks to demonstrate a nuclear thermal propulsion (NTP) system on orbit. (Credit: DARPA)

## CURRENT STATE

Despite lab-scale technical progress in the last year, there is still uncertainty on the DoD-side about the mission pull for next generation propulsion and power technologies and how exactly they will integrate with future satellite constellations and Cislunar infrastructure. Industry has recognized the fundamental nature of propulsion and power for spacecraft and the connection to revenue models. However, the elephants in the room are how SpaceX Starship will change access to Earth-orbit,[279] and how in-space refueling will be employed and compete or complement advanced propulsion systems.

**Progress on Nuclear Thermal Propulsion (NTP)** - Military and Civil NTP programs have progressed in the last year, with the Defense Advanced Research Projects Agency (DARPA) NTP program completing Phase I efforts and releasing a Phase II request for proposal[280] for more hardware

---

[278] United States Space Force (2020). Space Capstone Publication (SCP) Spacepower: Doctrine for Space Forces.
[279] Becker, J. (2021). A Starcruiser for Space Force: Thinking Through the Imminent Transformation of Spacepower. War on the Rocks.
[280] Erwin, S. (2022). DARPA moving forward with development of nuclear powered spacecraft. SpaceNews.

focused work, while NASA continues Phase I research on their NTP program.²⁸¹ These two government programs seem to share many technical challenges and are collaborating when and where appropriate. The DARPA NTP program expects to launch a demonstration vehicle by 2026,²⁸² while NASA's higher performance NTP concept may be demonstrated towards the end of the decade, benefiting from the regulatory, safety and licensing pathway forged by the DARPA program. These two programs are expected to make major headway on remaining technical challenges associated with NTP (e.g., high temp materials, long duration cryogenic storage) in the next several years, possibly opening up new commercial opportunities in nuclear thermal propulsion.

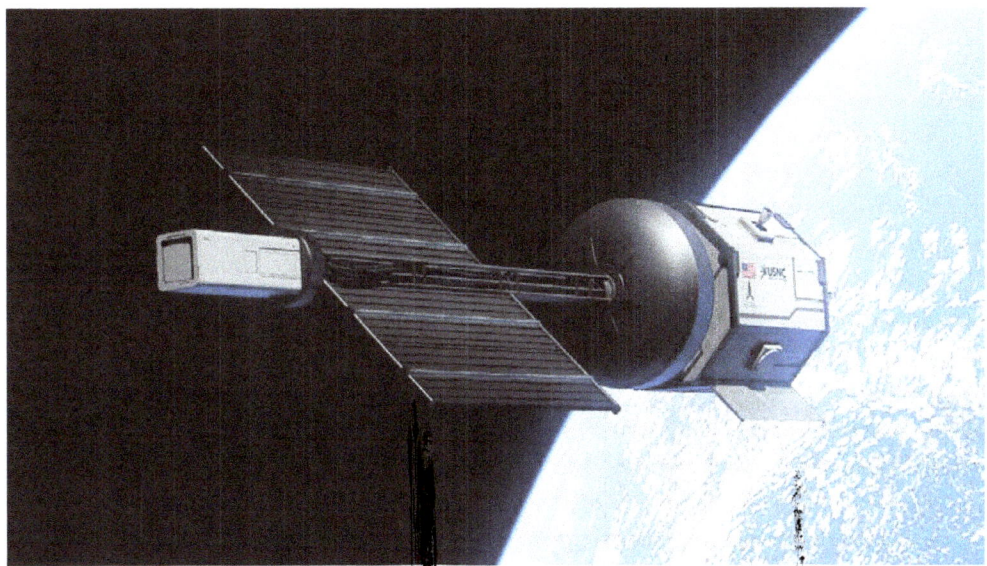

Figure 42. Artist rendering of the Shadow Runner spacecraft design utilizing 'next-gen' high power radioisotope source. (Credit: USNC-Tech)

**Fusion, Radioisotope, and Alternative Fuels** - The majority of research money in fusion is still focused on terrestrial power generation applications, with support from the Advanced Research Projects Agency-Energy (ARPA-E) and venture capital continuing over the last year.²⁸³ The White House OSTP hosted a summit to accelerate commercial fusion reactors with the goal of developing a carbon-free clean energy source.²⁸⁴ Recently, DIU awarded the first nuclear fusion DoD contract focusing on compact hybrid reactors for small satellites.²⁸⁵ Growing interest in alternative radioisotope materials (non-Plutonium) is evident with several DoD contracts (e.g., AFWERX, DIU),²⁸⁶ although guidance from the FAA is still pending on how launch licensing of radioisotopes will work for commercial companies. Technical and bureaucratic inertia from legacy fuels (e.g., hydrazine) are slowing adoption of green and non-toxic propellants, despite support from DoD stakeholders.²⁸⁷

---

[281] NASA (2021). NASA Announces Nuclear Thermal Propulsion Reactor Concept Awards.
[282] Erwin, S. (2022). DARPA moving forward with development of nuclear powered spacecraft. SpaceNews.
[283] ARPA-E (2020). Department of Energy Announces $29 Million in Fusion Energy Technology Development; Reed, S. Nuclear Fusion Edges Toward the Mainstream. New York Times.
[284] White House (2022). Readout of the White House Summit on Developing a Bold Decadal Vision for Commercial Fusion Energy.
[285] Defense Daily (2022). Contract Award: Zeno Power Systems Inc. (St. Louis, Missouri). Defense Daily.
[286] DIU (2022). DIU Launching Next-Generation Nuclear Propulsion and Power.
[287] AFRL (2022). Advanced Spacecraft Energetic Non-Toxic (ASCENT) Propellant.

**Power Beaming and Space Based Solar Power** - Commercial Photovoltaics (PV) are seeing interest in very high power arrays, possibly exceeding Megawatt (MW)- scale power output. Deployment and/or assembly of MW-scale solar arrays and associated technologies (e.g. high-voltage power processing) will enable space-based power beaming, but large uncertainty exists around the business case and demand signal for space-to-Earth beamed power. Technical progress has been made on converting solar to RF on the meter-scale and transmitting this power in the lab (AFRL SPIDR), although the U.S. is still lagging behind Chinese efforts towards large scale SBSP demonstrations, with the completion of a 75-m test tower at Xidian University this year.[288] These space-based

Figure 43. AFRL's ARACHNE will serve as the first free-flying flight experiment needed to mature critical technologies deemed essential to building an operational solar power transmission system. (Credit: AFRL)[289]

capabilities are predicted to have major military and economic impacts in the second half of this century[290] (with at least one market forecast so bullish as to suggest a market of $23B as early as 2040).[291]

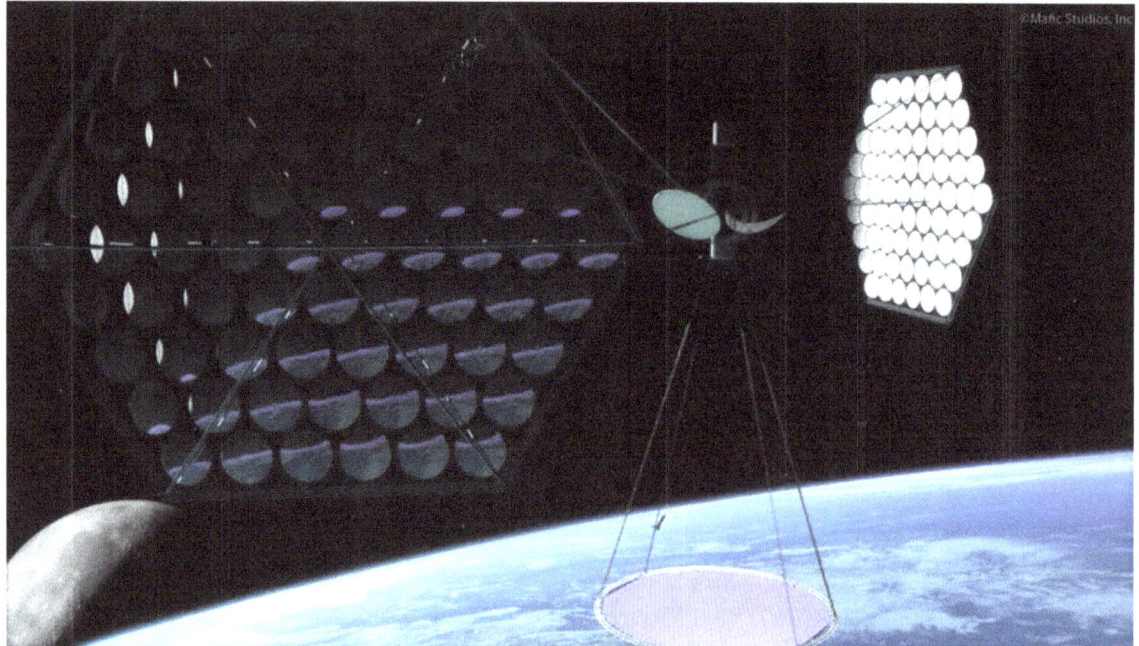

Figure 44. Rendering of a future space solar power system capable of harnessing the Sun's tremendous energy and beaming it to the surface of Earth as RF energy. (Credit: Kriss Holland ©Mafic Studios, Inc.)

[288] Jones, A. (2022). Chinese university completes space-based solar power ground test facility. SpaceNews.
[289] AFRL (2022). Space Solar Power Incremental Demonstrations And Research Project (SSPIDR).
[290] RAND (2022). Future Uses of Space to 2050; Frazer-Nash (2021). Space Based Solar Power: De-risking the pathway to Net Zero; Frazer-Nash (2022). Space-Based Solar Power: A Future Source of Energy For Europe?; Roland Berger (2022). Space-basedsolar power: Can it help to decarbonize Europe and make it more energy resilient?
[291] CitiGroup (2022). SPACE: The Dawn of a New Age.

> *"We forecast that the highest growth in the market will be from new space applications and industries, which will be unlocked as the industry becomes more affordable, accessibility becomes more widespread, and technology improves. This includes areas such as space-based solar power, Moon/asteroid mining, space logistics/cargo, space tourism, inter-city rocket travel, and microgravity R&D and construction."* -- CITIGROUP[292]

## KEY ISSUES & CHALLENGES

The largest obstacles to adoption of next generation propulsion and power technologies are lack of foresight and short funding horizons, regulatory and licensing hurdles, and risk aversion on all sides. In general, the government is not doing a good job of seeking out and supporting enabling technologies. In fact, government barriers slow or inhibit adoption of new technologies like power beaming and nuclear propulsion and power by requiring years-long regulatory licensing pathways.

**Mission Pull/Demand Signal** - To get away from the 'one-off' solutions that are typical in nuclear systems (e.g., Plutonium radioisotope power systems), industry needs a clear demand signal for how next-generation propulsion and power will integrate into future Cislunar posture. Participants lamented the apparent lack of curiosity from the government on what technologies are possible and how they might be used for economic, scientific or military advantage in the near future.

**Regulatory and Licensing Barriers** - Current timeline for a frequency allocation is several years, nuclear launch licensing procedures are likely longer, and specific commercial guidance does not exist. These long and painful processes indicate to industry that the government does not prioritize advancing state of the art in-space capabilities. In some cases, commercial companies are stuck in a 'Chicken and Egg' process across multiple regulatory bodies.

**Lost in the Valley of Death** - Current small business contracting process (e.g., SBIR/STTR[293]) differ from one agency to another, putting a large burden on companies to be able to respond to requests for proposals with innovative technology solutions. Lack of communication between DoD 'end users,' national labs, academia, and small business means promising tech dies in the lab before getting to an in-space prototype.

**Messaging the Costs & Benefits** - The return on investment in science, economic development and national security are significantly impacted by the decision to either use or not use safe sources of nuclear power and propulsion today and into the future (See Table 2).

---

[292] CitiGroup (2022). SPACE: The Dawn of a New Age.
[293] Air Force (2022). Program Overview; Air Force Small Business SBIR/STTR; NASA (2022). NASA SBIR.

| | WITH NUCLEAR POWER & PROPULSION |
|---|---|
| **Easier to Develop a Sustainable Cislunar Economy** | Fission surface power can provide multi-kilowatt power levels continuously in harsh environments for long periods:<br>• A 40kWe unit could provide sufficient power for a crewed habitat, in-situ resource utilization (ISRU), and large-scale exploration of the surface of the Moon.<br>• ISRU-derived propellant can enable exploration of the solar system. |
| | **WITHOUT NUCLEAR POWER & PROPULSION** |
| **Lower Return on Investment** | Mission Descoping/rescoping:<br>• Solar Probe Plus: Changes to avoid RTGs moved the mission further from the sun (fewer measurements possible) and reduced mission lifetime.<br>Poorer Science:<br>• Huygens (cost: >$660 million) - Had no more than three hours of battery life, most of which was planned to be used during the descent. Planners expected to get at most 30 minutes of data from the surface.<br>• Philae lander (cost: >€220 million) - Went into hibernation after just 60 hours of operation after landing on the comet 67P/C-G, and then did not operate again.<br>• (Future) Europa Lander (cost: UKN) - Doubtful that Lander could last more than 3 or 4 hours w/o nuclear power. |
| **Reduced Speed of Science** | New Horizon whizzed past Pluto:<br>• No ability to go into orbit.<br>• Spacecraft transmitted data at the rate of at most 1 kbps<br>• A full 16 months before all the data from the New Horizons' flyby of Pluto was received |
| **Greater Risk for Human Exploration** | Slower trip times to Mars:<br>• Higher exposure to galactic cosmic radiation and zero-g effects<br>• Order of magnitude lower number of launch windows<br>• Fewer en route abort options |

Table 2. Hard choices must be made to remain competitive in the economic development and human settlement of space (Credit: NASA)[294].

*"If you stop and think about it, the form of propulsion used today hasn't changed in over a thousand years... since the invention of fireworks by the Chinese. Basically you burn (oxidize) a material in a tube, hot gasses come out one end and the vehicle flies in the opposite direction. Sure our rockets have gotten bigger and more efficient, but the basic design remains unchanged."*

– DR. PETER DIAMANDIS, 2010[295]

---

[294] Lal, B. (2022). Keynote Presentation: Going Nuclear - Prospects for Power and Propulsion in Space. SSIB'22 Workshop.
[295] Diamandis, P. (2010). Personal Spaceflight Industry - Reflections on the Five Year Anniversary of the Ansari X PRIZE. Huffington Post.

Figure 45. NASA's Kilopower fission reactor seeks to provide 10 kilowatts of power necessary to sustain life and economic activity on the Lunar surface. (Credit: NASA)

## Nuclear Power is Pivotal to Human Settlements on Moon & Mars

This time when humankind returns to the Moon, it intends to stay by establishing a permanent station capable of supporting life, exploration and economic development activities.[296] Such a feat is not possible without reliance on a safe and reliable energy source capable of supplying tens of kilowatts of continuous electrical power for years. Although the sun provides kilowatts of power to the International Space Station, the Moon is gravitationally locked with its massive companion (the Earth). As a result, the Lunar night is two Earth weeks of total darkness with an average temperature of -280°F or lower.[297] Solar arrays would be shadowed by the Moon; therefore, nuclear power is the only viable solution. But is it safe? The answer is YES.

Nuclear power has been used in space for more than 50 years. The SNAP-27 was a radioisotope thermoelectric generator deployed on the Moon during the Apollo era. It provided 70 watts of continuous electrical power to the Apollo Lunar Surface Experiment Package (ALSEP) during both Lunar day and night conditions.[298] The SNAP-27 was designed to work for 2 years, but reliably worked for 8 years until NASA finally turned it off. A broad selection of nuclear power options are available today providing even greater safety and reliability. Kilopower's U-235 core carries fewer than 5 curies of total radioactivity while inert on the launch pad. By comparison, a hospital's radiotherapy machine might contain as much as 1,000 curies of a potent radioisotope.[299]

Not surprisingly, some of the most interesting new innovations in nuclear power (and propulsion) are coming from the commercial sector. Some employ radioisotopic fuels with short half-lives capable of powering portable systems and small satellites, which would otherwise have low duty cycles between battery recharging. Others will be capable of providing significant thrust and power for human missions to Mars and beyond. The quest for safe fusion and other reactors will not only improve life in space, but are essential to enable humanity's evolution from fossil fuel energy sources harmful to Earth's unique biosphere. In space, power is life; and reliable continuous sources of power will prove pivotal to building and sustaining an off-planet economy and future human settlements.

---

[296] Choudhury, S. (2022). We Can Already Build a Moon Base, So Why Haven't We Yet? The Medium.
[297] NASA (2022). NASA's LRO Finds Lunar Pits Harbor Comfortable Temperatures.
[298] Barklay, C. et al. (2021). Can MMRTG Operate on the Moon? Insights from SNAP-27 for Apollo Lunar Surface Experiment Package. IEEE. DOI: 10.1109/AERO50100.2021.9438465.
[299] Jones, T. (2019). Space nuclear power - seriously. Aerospace America.

## KEY INFLECTION POINTS

**Nuclear Launch Success or Incident** - a reactor or radioisotope power source is launched successfully. Alternatively, an explosion on the pad or in flight, causes a dispersal of radiological materials - would be a major setback to public perception of nuclear power in space.

**Starship reaches orbit demonstrating 100-ton to LEO** - would be a stepwise change in launch cost and access to space. The impacts to commercial business plans and how this will enable new concepts of operation in Earth orbit are still being understood. A successful Starship orbital demo may also inform feasibility of high-mass space based solar power (SBSP) beaming projects and MW-scale nuclear fission reactor space propulsion and power concepts.

**Fusion ignition** - a successful demonstration of a reactor that produces more power than it uses (Q>1) would be an important milestone for terrestrial power generation and spark further investment and acceleration of fusion applications in space. Several companies and a national laboratory claim to be on the brink of this breakthrough.[300]

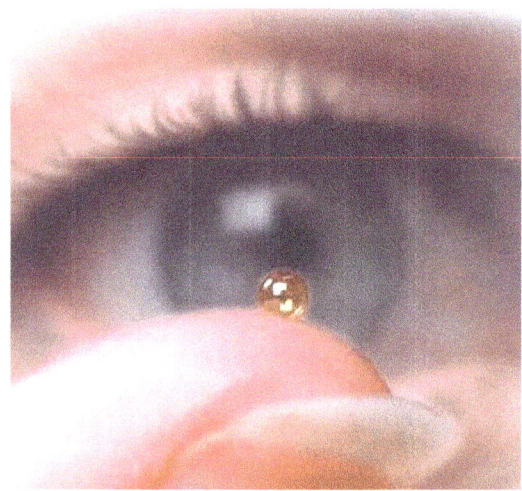

Figure 46. Chilled close to absolute zero, this microcapsule, filled with deuterium-tritium ice, released more energy than it absorbed from the National Ignition Facility's bank of 192 lasers last year. (Credit: LLNL)[301]

**Megawatt-scale space to ground demo** - would illustrate how to successfully navigate technical, regulatory, and licensing hurdles, while bringing together elements of in-space servicing and manufacturing (ISAM) and proving that SBSP can be scaled enough to develop a new market for terrestrial energy.

## KEY ACTIONS & RECOMMENDATIONS

### SHORT-TERM PAYOFF

**Shepherd not Sherpa:** The government needs to 'show' not 'carry' small businesses through regulatory, licensing, contracting, and proposals submission processes. This increases gov access to innovative technologies and lowers barriers for small business. (OPRs: SBA, DoD, IC, NASA)

**Assess and develop a military use case analysis:** Create a clear demand signal and associated joint requirements document for advanced and/or breakthrough propulsion and power concepts, including development of innovative concept of operations in space (OPRs: USSF, DIU, Strategic Capabilities Office (SCO), AFRL, DARPA, Space Development Agency (SDA) Space Rapid Capabilities Office (SpRCO))

---

[300] Thomson J. (2022). Nuclear Fusion Breakthrough Confirmed: California Team Achieved Ignition. Newsweek.
[301] Kramer, D. (2021). Lawrence Livermore claims a milestone in laser fusion. PhysicsToday.

Figure 47. President Kennedy addressing a joint session of Congress in 1962. (Credit: JFK Library)

*"First, I believe that this nation should commit itself to achieving the goal, before this decade is out, of landing a man on the moon and returning him safely to the Earth...*

*Secondly, an additional 23 million dollars, together with 7 million dollars already available, will accelerate development of the Rover nuclear rocket. This gives promise of someday providing a means for even more exciting and ambitious exploration of space, perhaps beyond the moon, perhaps to the very end of the solar system itself."*

- PRESIDENT JOHN F. KENNEDY, 1962[302]

**Small Business Innovative Research (SBIR) Phase 1.5** - Support companies building disruptive technologies that are stuck in between Phase 1 and Phase II[303] technology readiness level with more opportunities for 'intermediate' level awards. This increases government access to innovative technologies and increases chances of a successful prototype or capability. (OPRs: SBA, DoD, NASA)

**Government-sponsored space-based testing facility** for advanced and breakthrough propulsion and power research, testing, and evaluation. Many nuclear technologies present a challenge for testing on Earth. A government sponsored orbital 'user-facility' could help transition prototypes and advance technology readiness levels. (OPRs: USSF, SDA)

## LONG-TERM PAYOFF

**Accept more Risk on low TRL (<3) propulsion and power tech** - Perception from industry is that government programs are still unwilling to take any risks on new space technologies. U.S. led efforts to support early-stage research and development of high-risk high-reward space technologies will have long-term strategic payoffs. (OPRs: DoD, NASA, OSD R&E)

**Homestead Act for Space** - the U.S. government could provide services or materials in orbit and guarantee free or fixed cost for users. In the case of advanced power and propulsion, examples are fuels (e.g., xenon), chargeable atomic batteries and beamed power. Providing these materials or services could help accelerate adoption of the technology. (OPRs: OSTP, NSC)

---

[302] White House (1961). President Kennedy: Special Message to the Congress on Urgent National Needs. Delivered in person before a joint session of Congress on May 25, 1961. JFK Library.
[303] Air Force (2022). Phase I and II.

Figure 48. BlackSky's small satellites provide high resolution, electro-optical imaging to support customers including Ukrainian Defense Forces in countering Russia's unprovoked invasion. (Credit: BlackSky)

# REMOTE SENSING & TRAFFIC MANAGEMENT

*"This conflict has brought it to the forefront where people are realizing, 'Wow, what an advantage these commercial space remote sensing capabilities provide.' I think that this is changing the way that we fight [and] the way that the world is able to see the fight."*

— TODD HARRISON, 2022[304]
Former Director, CSIS Aerospace Security Project

## BACKGROUND

Remote sensing has been a focus of prior SSIB reports, however space traffic management is a new topic for SSIB'22. The combined topics cover a wide range of technical areas, and this working group was chartered to review the current remote sensing and traffic management landscape, and to propose implementable recommendations needed to resolve undetermined aspects of operation in Cislunar space, spectrum allocation, space situational awareness, and Earth observation with real time tracking from, to, and in space. The group also held discussions on modernizing and overcoming barriers to commercial acquisition and leverage the nation's growing commercial remote sensing industry to enhance our space domain awareness (SDA) architecture and ongoing expansion of LEO constellations.

Given the breadth and depth of these topics, we conducted a survey to prioritize and focus the discussion to solicit the feedback from the industry team attendees. The results of the survey with 42 participants on the relative importance are shown to the right. The top three topics as voiced by respondents to the survey were the real-time tracking from, to, and in space; Cislunar space and space situational awareness, and leveraging the nation's growing commercial remote sensing industry. This set reflects the group's consistent voice that the attention and resources paid to Earth observation capabilities over the last several years must also be matched by advocacy for leveraging the same industry to address gaps in coverage for situational awareness, particularly at extended orbits and regions.

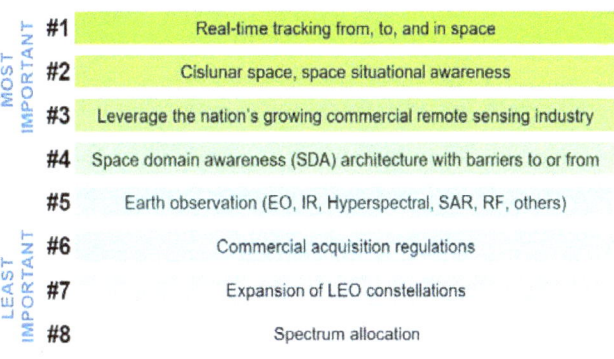

Figure 49. Results from Working Group Survey of 42 Respondents. (Source: NSNM)[305]

Though this is the first year for this working group to consider SDA, the consistent theme from the industry was that we are not making enough progress, Space traffic management has no real funding and no viable advocate within the U.S. government. However, commercial remote sensing has made significant strides in operationalizing commercial space capabilities, as demonstrated in the

---

[304] Feldscher, J. (2022). The Ukraine War Is Giving Commercial Space an 'Internet Moment'. DefenseOne.
[305] New Space New Mexico (2022). Survey of Remote Sensing and Traffic Management Working Group Attendees.

Russo-Ukraine conflict, and The National Reconnaissance Agency's (NRO) award to Maxar Technologies, Planet, and BlackSky contracts[306] for commercial remote sensing imagery, each with a five-year commitment and options stretching to 2032.

Figure 50. Satellite imagery taken by Maxar on February 22, 2022, showing military buildup occurring at V.D. Bolshoy Bokov Airfield, Mazyr, Belarus just 37 37 km north of the Ukraine border. (Credit: Maxar)[307]

## CURRENT STATE

**Commercial Remote Sensing for the Global Good** - the Russian invasion of Ukraine has solidified the need, relevance, and value of commercial remote sensing as a means of transparently vetting geopolitical messaging against reality.[308] It has brought glaring attention to the capability of commercial remote sensing companies to provide a timely record of war events and casualties, and the economic consequences flowing from the same. Commercial remote sensing data therefore strengthens the "Information" element of U.S. National Power by further enabling detection and subsequent prediction of military maneuvers, and then distributing that information to people across the globe who may be contending with misinformation and disinformation. In fact, this has led to an emergence of commercial companies that are reporting disasters, both natural and human caused, to the public faster than governments can form state messaging.[309] Increasing availability and responsiveness of commercial sensor data will continue to improve its own potency as a form of "open-source deterrence." There was no question among the working group attendees that the commercial business for remote sensing extends well beyond the DoD to other civil consumers such as the Department of

---

[306] Davenport, C. (2022). US NRO awards billions to commercial satellite imaging companies. Washington Post.
[307] Wood, S. (2022). Keynote presentation: Russia's Invasion of Ukraine: Providing Transparency and Truth With Commercial Satellite Imagery. SSIB'22 Workshop.
[308] Wolfe, F. (2022). Russian Assault on Ukraine Highlights Potential of Commercial Satellite Imagery to Predict/Head Off Conflicts. Defense Daily.
[309] Mayday (2022). Services – Mayday.ai.

Agriculture, Department of State, FEMA, Environmental Protection Agency, NASA, NOAA, U.S. Coast Guard, and other private and commercial users such as the news media industry.[310]

Figure 51. NASA Harvest image using Planet's data shows fields in the Russian-occupied area north of Melitopol, Ukraine. (Credit: Planet)[311]

**The State of Space Domain Awareness and Space Traffic Management** - Space Domain Awareness (SDA) and Space Traffic Management (STM) are discrete but related missions and so share common language. USSF's SDA mission, sometimes referred to as "catalog maintenance" or the now-deprecated term "space situational awareness," is to protect and defend U.S. interests in space, which certainly relies on knowing where things are in orbit at any moment. Space Traffic Management is broader in scope and is not specific to the protect and defend mission, but rather enables the safe and good order of all activities in space. The working group notes an unresolved question over the future of Space Traffic Management (STM), particularly on the matter of who will hold executive agency over the mission – the DoD (USSPACECOM), or a civil body (Department of Commerce or the FAA). There is not yet any strong indication from within the Executive of Legislative Branches about which will happen, but the working group recommends that DoD/USSF retain leadership and authority for commercial remote sensing and space domain awareness activities, and that the Department of Commerce be established as lead for Space Traffic Management.

**A Mixed Bag of Positive Outcomes with Growing Concerns** - The last 12 months has demonstrated both progress and stagnation. NRO's award of multiple contracts to remote sensing companies is a positive step toward providing meaningful adoption of these services, while some looming concerns over the level of commitment and consistency of that adoption, licensing and distribution restrictions, and supply chain persist. This "mixed bag" sets the stage for a number of the inflection points detailed further below, but also presents opportunities to set the U.S. and the space industrial base on a stronger course for success and increasing prosperity.

---

[310] USGS (2022). Commercial Remote Sensing Space Policy.
[311] Parks, S. (2021). Leveraging Planet Satellite Imagery To Improve Irrigation Intelligence. Planet.

## KEY ISSUES & CHALLENGES

Issues and challenges discussed by the working group centered primarily around licensing and associated policy, concerns over the U.S.' ability to maintain Space Domain Awareness (SDA) and safely execute space traffic management, and the extent by which the U.S. Government and Department of Defense leverage commercial remote sensing data domestically and abroad. The working group agrees that addressing these issues and challenges requires a mixture of policy, procurement, and collaborative government-commercial leadership to overcome.

**Guns, Germs, and Steel** - Paying homage to Jared Diamond's 1997 book,[312] the last 12 months are a confluence of the emergence of war in Ukraine, the ongoing COVID-19 pandemic, and global markets that consequently remain under enormous stress. Commercial imagery has proven invaluable in the Ukraine conflict, both to the U.S. government's efforts to counter the Russian information war, as well as to the Ukrainian military.[313] The multi-year, multi-billion awards by NRO are certainly movements in the positive direction for the EO and SAR commercial marketplace. However, additional challenges both due to current policies and COVID-19 related supply chain woes continue to impact the space industrial base.[314] Concerns about long-term business cases for commercial remote sensing, the lack of meaningful policies that allow what is technically feasible, and uncertain funding for space traffic management also continues to negatively impact the industry. Specifically, it is the manufacturing sector that has had the greatest impact due to the supply chain issues. Though the presence of manufacturing industry performers was limited, those present have been unable to build what they had planned to, and current policies are prohibiting them from building to the capabilities that are technically feasible now.[315]

Figure 52. Dawn-to-dusk images collected over Korczowa-Krakovets border show progressive increase of vehicle traffic leaving Ukraine over a nine-hour period on March 17, 2022. (Credit: BlackSky)

**Licensing, Licensing, Licensing** - Participants in the working group reported little meaningful progress on licensing, and stated they face significant difficulty in competing with international remote sensing companies who are not bound to the same restrictions imposed in the U.S.. Further, U.S. companies are subjected to additional strategic risk by developing technologies that can't be exported, when international competition doesn't face the same hurdles. One exception where notable progress has been made is in the ability for U.S. companies to conduct select satellite-satellite imaging activities.

---

[312] Diamond, J. (1997). *Guns, Germs, and Steel*. W. W. Norton & Company.
[313] Oxendine, C. (2022). Ukraine: Open-Source Data Aided Response and Documents Damages and Atrocities. ESRI.
[314] Magableh, G. (2021). Supply Chains and the COVID-19 Pandemic: A Comprehensive Framework - PMC. NIH.
[315] Domestic imagers are currently prohibited from selling imagery with resolution better than 50 cm in panchromatic, or black and white, two meters in the multi-spectral and 7.5 meters in shortwave infrared (IR), according to DigitalGlobe [DGI] Executive Vice President and Chief Technology Officer (CTO) Walter Scott.

**U.S. Space Surveillance Lacks Coverage** - There remains a significant lack of coverage, both in terms of the number of sensors and the orbital diversity and geometry of those sensors. Under the mantra of "every object, every orbit," comprehensive SDA means responsive detection, characterization, and custody in order to update the catalog and dependent models faster than changes occur. With gaps in coverage for the current Space Surveillance Network (SSN),[316] specifically in LEO and more increasingly in GEO and eventually beyond GEO, the U.S. is not postured to adequately execute intelligence and space traffic management missions. In particular, if current trends continue, LEO satellite numbers may grow as high as 50,000 satellites over the next 10 years,[317] many of which will only be detected infrequently by the SSN. The "demand" for STM will soon exceed supply of sensors, so it is imperative that the U.S. invest in commercial SSA.

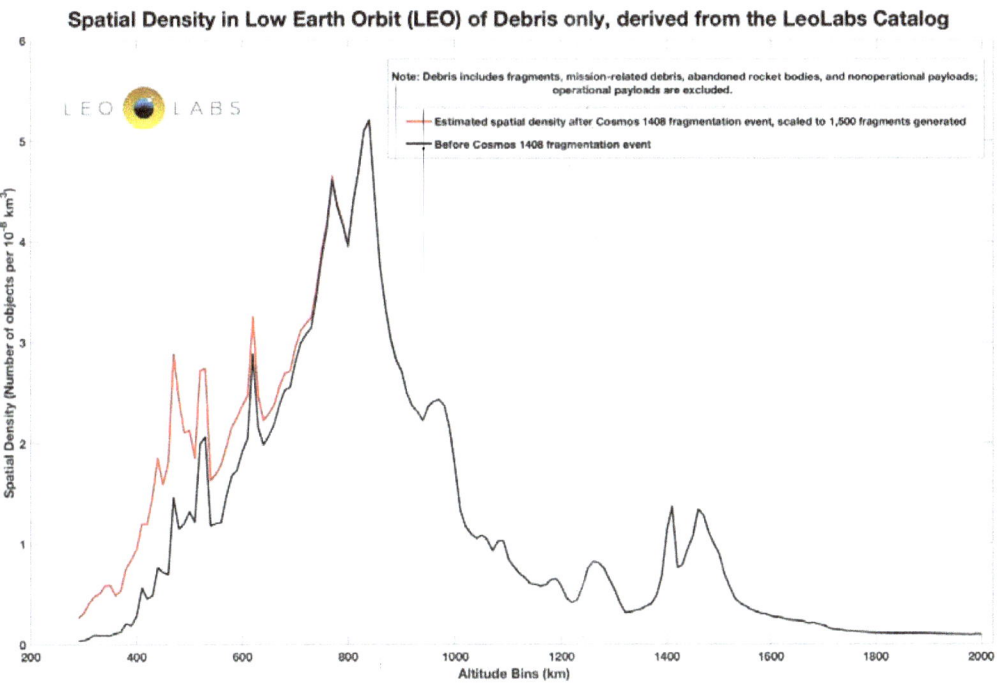

Figure 53. Changes in orbital debris density resulting from the Russian ASAT test and corresponding breakup of the Cosmos 1408 satellite. Some of the new Cosmos 1408 debris fragments may be as large or larger than many cubesats operating in LEO. (Credit: LeoLabs)

**Traffic Management for a New Age** - The important task of space traffic management must be re-framed against the constellations of today, and both their accelerating growth and increasing maneuverability. Without addressing the space domain awareness challenge identified above, and without allocating a budget in line with the scale of this need, the U.S. cannot effectively sustain its leadership in space traffic management across all altitudes and regions. As the proliferated constellations become the new normal, industry is likely to buy data services from commercial providers to maintain their situational awareness on a timely basis and therefore grow a commercial STM infrastructure with more than just U.S. government funds.

---

[316] White House (2018). Space Policy Directive-3, National Space Traffic Management Policy.
[317] Daehnick, C., Klinghoffer, I. Maritz, B. & B. Wiseman (2020). Large LEO satellite constellations: Will it be different this time?. McKinsey.

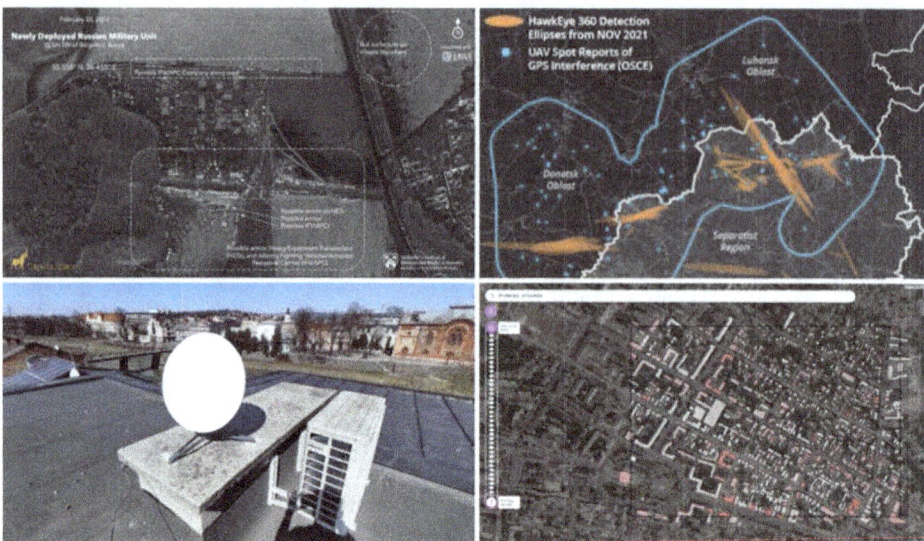

Figure 54. Commercial solutions aiding Ukraine in its defense against Putin's invasion.
(Sources clockwise: Capella Space, Hawkeye 360, DIU/Microsoft, Starlink)

## Commercial Tech has forever changed the Character of Conflict

When Russia invaded Ukraine in 2014, the unexpected events led one BBC reporter to observe, "the annexation of Crimea was the smoothest invasion of modern times. It was over before the outside world realized it had even started."[318] Eight years later, commercial space technology has profoundly changed the character of conflict. Months before this February's invasion of Ukraine, commercial remote sensing satellites provided early indications and warning of Vladmir Putin's sinister intentions. Armored columns literally had no place to hide regardless of day, night or inclement weather conditions.[319] Unclassified images were readily accessible by government leaders, the media and the public providing Ukrainians valuable time to prepare.

Once the conflict was underway, commercial satellite communications became the lifeline for disbursed tactical units to maintain situational awareness while sharing video and digital images of Russian atrocities for all the world to see. At the strategic level, President Zelenskyy's ability to communicate with the outside world via commercial satellite communications was never interrupted. Commercial cloud computing platforms processed vast amounts of open source data and imagery using computer vision algorithms to identify bomb damage to churches, schools and civilian housing. Hence, the United Nations has been able to accurately track the scale of destruction with daily precision.[320] Russian interference of GPS signals over land and sea was detectable using commercial small satellites optimized for detection in the radio frequency spectrum. The agile nature of commercial space innovation has significantly aided a small nation in its defense, while slowing the offensive of a great regional power. These and other developments have certainly given China cause for concern as it considers its hegemonic ambitions vis-à-vis Taiwan.

Now is the time to embrace commercial space capabilities to meet the needs of friends and allies around the world. The Belt and Road Initiative has long sought to displace western sources of satellite imagery, communications and social media. Imagine a world where autocracies like China and Russia control such tools with the ability to shape the truth as they see it. Commercial space technology has proven its value. It should be acquired at scale to ensure its value well into the future.

---

[318] Simpson, J. (2014). Russia's Crimea plan detailed, secret and successful. BBC.
[319] Aldhous, P. and Miller, C. (2022). How Open-Source Intelligence Is Helping Clear The Fog Of War In Ukraine. Buzzfeed.
[320] Schroeder, A. (2022). Analyzing the Scale of Ukraine's Destruction. ReliefWeb.

**Mandating Commercial Utilization** - The U.S. Government's use of commercial has not yet been institutionalized for procurement of remote sensing data. It is still hamstrung by its inability to solicit for solutions (to include services) instead of specific technologies. It isn't making enough bets, and isn't adequately resourcing the commercially-oriented and OTA-equipped DoD offices that are able to move faster to achieve a higher transition rate than more traditional pathways.

## KEY INFLECTION POINTS

**Commercial sensors attain 5 to 10 cm resolution, and constellation density such that 90+% of data is available from these sources** - For most DoD operational needs, the national systems become 'functionally' irrelevant to the vast majority of consumers/users, particularly at the tactical level.[321]

**Commercial Remote Sensing (with exception of Space Domain Awareness) is fully realized** - alongside established service procurement pathways as an integral component of the U.S. and allied information collection and exploitation frameworks.

**A U.S. company develops a breakthrough technology and is able to bring it to market faster abroad than within the U.S.** - due to continued licensing constraints and export controls.

**China establishes dominance in the global remote sensing market** - by number of sensors, and is quickly adding customers through simple purchasing mechanisms. The U.S. retains the crown for fidelity and data quality, but at higher prices and with USG/ITAR-imposed restrictions intact.

Figure 55. Capella's next generation Acadia synthetic aperture radar (SAR) satellite with 21 cm (slant range) to 31 cm (ground range) resolution and optical inter-satellite links that are compatible with the Space Development Agency's National Defense Space Architecture. (Credit: Capella Space)[322]

---

[321] Nuclear Communications, Command & Control (NC3), IC and DoD will and should maintain the core capabilities.
[322] D. Werner (2022). Capella unveils a new generation of radar satellites. SpaceNews.

**Space Force becomes an executive agent for commercial space services** - and can procure remote sensing data (and SDA) for all DoD users, without having to go through the intelligence community.

## KEY ACTIONS & RECOMMENDATIONS

These recommendations reflect the numerous inputs from workshop participants and their best assessment regarding which agency(cies) are in the best position to forward the necessary change.

### SHORT-TERM PAYOFF

**Mandate the purchase of commercial solutions when available** and maximize the use of Commercial Solutions Openings (CSO) and similar OTA-based processes to solicit solutions to remote sensing and traffic management problems. In doing so, DoD will better communicate its needs, and empower industry to bring dual-use solutions to the table. (OPRs: DoD, USD A&S, USD R&E)

**Resource DoD (with authorities, procurement methods, and funding) to buy data services** for improved SDA and space traffic management missions. (OPR: Congress)

**Establish USSF as the Executive Agent for Commercial Remote Sensing and Space Domain Awareness** at the unclassified and collateral secret level, and have Dept of Commerce lead the effort to explore government options and facilitate collaboration between private stakeholders' to conceive, develop, field and operate an effective space traffic management architecture (OPRs: Congress, DoD, USSF, DoC)

### MID-TERM PAYOFF

**Enact licensing improvements for commercial imagery** to keep U.S. companies competitive in the international marketplace. This entails the release of full-resolution imagery, continued support for commercial satellite-to-satellite imaging, and expanding releasability of high resolution commercial imagery. (OPRs: Congress, OSTP, DoC, DoD)

**Designate the Department of Commerce to lead an interagency program office overseeing all space traffic management activities** - The U.S. must retain leadership in this role. (OPRs: Congress, DoC coordination, ODNI, NASA, DoD)

**Mandate that at least 50% of remote sensing data is purchased from U.S./allied commercial providers within the next 5 years** taking advantage of the expected 100:1 ratio of commercial to government sensors. Exempting the agency when it can justify an exception to compliance, does not violate security rules and policies, such as a violation of the agency's own charter to provide a service like protect and defend. (OPRs: Congress, OSTP)

### LONG-TERM PAYOFF

**Increase the mandated requirement for purchase of commercial remote sensing data to 80%** in order to set conditions for the "GEOINT Singularity" (true and ubiquitous convergence of data and analytics for full-Earth coverage). (OPRs: OSTP, Congress, DoD, ODNI, NOAA)

**Develop infrastructure for public/private partnerships (with USSF)** in order to achieve persistent coverage from GEO and beyond for SDA. (OPRs: OSTP, Congress, DoD, NASA)

**Empower the Department of Commerce to lead globally** for the development and oversight of all space traffic management activities. (OPRs: OSTP, Congress, DoC)

**Grow the budget of the U.S. Government STM office in DoC to $300M over a 10 year period** - Then sustain at least that level. Establish and Provide U.S. leadership in STM as in commercial flight management, in collaboration and allies and in coordination with industry. (OPRs: Congress, DoC)

**Advocate and incentivize transponders and collision avoidance systems for spacecraft.** The Office of Space Commerce can advance safety by advocating for spacecraft and boosters to carry radio identification broadcasting tags, analogous to maritime AIS or aviation ADS-B, and on-board traffic collision avoidance systems, analogous to aviation's Traffic Collision Avoidance Systems (TCAS). (OPRs: DoC/OSC)

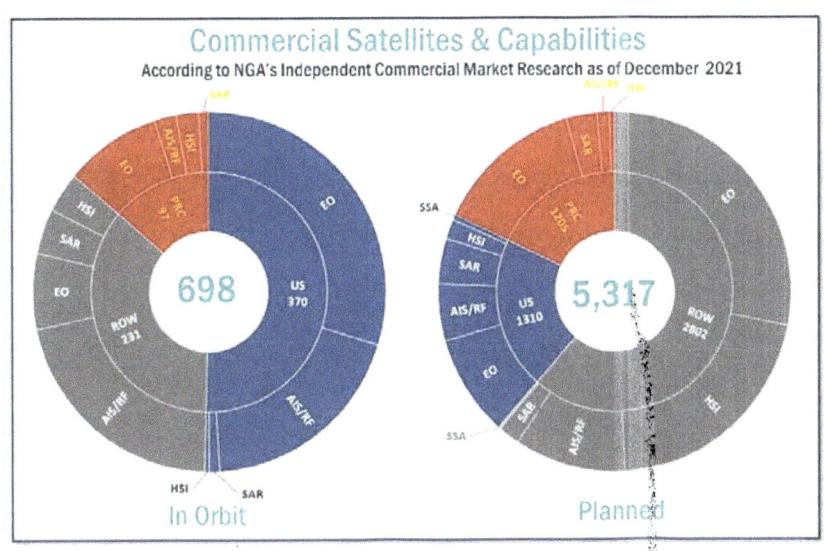

Figure 56. Planned growth in global commercial remote sensing. (Credit: NGA)

Figure 57. Vice President Kamala Harris speaking at NASA Goddard Space Flight Center on November 5, 2021. (Credit: Bloomberg/Licensed from Getty Images)

# POLICY & FINANCE

*"While SpaceX and others have been taking bold steps toward paving a technical path toward space settlement, there is an equivalent need to take bold steps toward enacting the necessary policies that will enable the expansion of human civilization into the solar system and beyond."*
— DR. SIMON "PETE" WORDEN, (Brig. General, USAF, Ret.)

## BACKGROUND

Stakeholders in the national security space community highlight the inextricable link between the development of space and the advancement of critical domestic and national security policy priorities. The 2021 SSIB report articulates the key role space-based technologies play in combating climate change, powering financial transactions and the energy grid and providing navigation and timing for our military assets, among other key functions. However, there remains several non-trivial financial, infrastructure, and technological challenges to ensuring U.S. competitiveness.

The Biden Administration and Congress are pursuing several policy priorities through both legislation and executive action to strengthen U.S. technological competitiveness, ensure sustained and inclusive domestic economic growth, strengthen the industrial base and build resiliency into supply chains, combat climate change, improve U.S. infrastructure, and shore up U.S. partnerships and alliances internationally. While not an exhaustive list, some specific actions and recommendations include:

- Bipartisan Infrastructure Law[323]
- Bipartisan Innovation Act[324]
- Build Back Better Framework[325]
- Executive Order on America's Supply Chains[326]
- Creation of the U.S. - EU Trade and Technology Council[327]
- Maintenance and Expansion of the Quad[328]
- Launch of AUKUS Grouping[329]
- National Climate Task Force[330]

---

[323] Department of Transportation (2022). Bipartisan Infrastructure Law | FTA.
[324] Congress (2021). S.1260 - 117th Congress (2021-2022): United States Innovation and Competition Act of 2021; White House (2022). Remarks By President Biden on the Bipartisan Innovation Act.
[325] White House (2021). The Build Back Better Framework.
[326] White House (2021). Executive Order on America's Supply Chains.
[327] U.S. Trade Representative (2021). US-EU Trade and Technology Council (TTC).
[328] The White House (2022). FACT SHEET: Quad Leaders' Tokyo Summit 2022.
[329] Office of the President of the United States (2021). Joint Leaders Statement on AUKUS. The White House.
[330] White House (2021). National Climate Task Force.

Stakeholders in the national security, civil, and commercial space sectors have an opportunity to engage and shape these various efforts to ensure that the critical underpinnings of the space industry are funded, that standards and norms for advanced technology reflect space requirements, and that there is a regulatory environment that is conducive to trade, investment and increased sharing of knowledge amongst partners and allies. It will be incumbent on policymakers, commercial actors, and academics to ensure that the equities of the space community are well-represented in this unique moment of economic and national security strategic planning.

Furthermore, with the Biden Administration's push for stronger international engagement through the launch of AUKUS, the expansion of the Quad, and ongoing support for Ukraine, the U.S. commercial space industry is poised to play a critical role.[331] Commercial technology is uniquely suited to strengthen relationships with international partners because it can be shared freely among partners without the complication of classification levels and strengthens interoperability. Additionally, commercial technology can be fielded in 1-2 years rather than 1-2 decades. However, increased reliance on the commercial sector raises a variety of foreign ownership, control, or influence (FOCI) concerns from the interconnected, global supply chains to adversarial investment to exports. To date, there is a lack of clarity about the expectations and methods of communication between the government and industry on these concerns. With growing international engagement, it is likely that areas of friction and misunderstanding will only widen.

---

*"The United States, you see, is the flag of choice for space activities. It is the flag of choice. And with that comes great opportunity but also great responsibility in terms of what course we will chart for the work that happens here on Earth that will then maximize the opportunities in space... To that end, we understand that we have got to update the rules, because they're just simply outdated. They were written for a space industry of the last century."*
— VICE PRESIDENT KAMALA HARRIS, 2022[332]

---

## CURRENT STATE

The U.S. space industry continues to grow as the commercial, civil and national security demands for services and experimentation increase. The global space economy continued to expand, as space startups raised nearly $47 billion in 2021, including $14.5 billion in the fourth quarter.[333] However, as inflation continues at record levels, owing in various parts to COVID induced supply chain bottlenecks, significant monetary and fiscal interventions during the height of COVID, the war in Ukraine, and its attendant effects on energy and food supplies, the Federal Reserve is ramping up its responses to cool prices. In response, the private funding environment is slowly beginning to retrench and the space

---

[331] Office of the President of the United States (2021). Joint Leaders Statement on AUKUS. The White House; Office of the President of the United States (2022). FACT SHEET: Quad Leaders' Tokyo Summit 2022. The White House; Office of the President of the United States (2022). Statement by President Joe Biden on Support for Ukraine and Call with President Zelenskyy of Ukraine. The White House; National Reconnaissance Office (2022). Press Release: NRO announces largest award of commercial imagery. NRO.

[332] White House (2022). Remarks by Vice President Harris On Supporting the Commercial Space Sector. The White House.

[333] Space Capital (2022). Space Capital Quarterly Report.

ecosystem will likely feel the second and third order effects of less funding or down-rounds in the coming months. Early data shows that space startups secured $7.2 billion in Q1 2022, with expectations that the second quarter figure will be significantly lower.[334]

The technologies and capabilities being developed in the space industrial base, however, remain more important than ever. Perhaps most notably, the conflict in Ukraine underscores the importance of dual-use space-based technologies, such as commercial satellite imagery, which is enabling real-time insight into Russian military activities. The reliable and multiple sources of truth are effectively countering attempts by the Kremlin to promote false narratives and providing significant ballast to Ukraine, the NATO alliance and the global community in turning back this unwanted aggression.[335] As the National Reconnaissance Office highlighted, "Due to its unclassified and shareable nature, commercial remote sensing data offer important benefits, including increased transparency, mission critical awareness, and humanitarian assistance."[336]

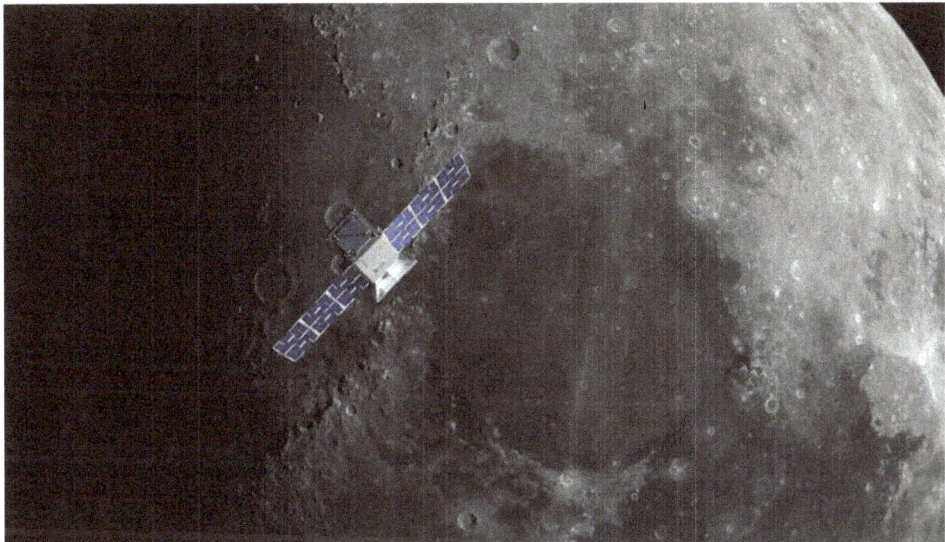

Figure 58. The Cislunar Autonomous Positioning System Technology Operations and Navigation Experiment (CAPSTONE) uses a commercial spacecraft to test the near-rectilinear halo orbit (NRHO) around the Moon that will be used by Artemis missions. (Credit: NASA)

Building on the momentum of the previous year, the Biden Administration continues to organize government efforts and push the frontiers of space exploration and exploitation. Most recently, the OSTP and the National Space Council released its In-Space Servicing, Assembly, and Manufacturing (ISAM) Strategy, outlining how the Administration will "support and stimulate the USG, academic, and commercial ISAM capability development."[337] Further, in the civil domain, NASA's novel partnership with commercial companies, Advanced Space and Rocket, among others, enabled the launch of the CAPSTONE spacecraft; a first step in the Artemis journey that is being executed at an

---

[334] Axios (2022 July 12). Axios Pro Rata.
[335] Albon, C (2022). How commercial space systems are changing the conflict in Ukraine. C4ISRnet.
[336] National Reconnaissance Office (2022), Press Release: NRO announces largest award of commercial imagery.
[337] Office of Space Commerce (2022). National Strategy for In-Space Servicing, Assembly, and Manufacturing Released. Department of Commerce.

efficient cost, on an expeditious schedule, and delivering a capability that is not exclusive to a government mission.

In addition to private sources of financing, the USG is responding to the call from industry to pursue creative financing vehicles to seed the space industrial base. In particular, the U.S. Export Import Bank launched the Make More In America Initiative which, when combined with ExIm's China & Transformational Exports Program (CTEP), is designed to unlock loans and loan guarantees to "deep tech" industries such as space, enabling U.S. companies access to low-cost capital in order to scale manufacturing in the U.S.[338] The USAF, through AFWERX and a specific initiative solely focused on space capabilities, SPACEWERX, continues to utilize its SBIR program to provide non-dilutive funding and create greater leverage with private investors.[339] And the recent announcement by Under Secretary of Defense for Research and Engineering Heidi Shyu on the creation of the Office of Strategic Capital at the Department of Defense, are all signals that the USG is working to develop a coherent way of providing non-trivial financial interventions in critical technology areas.[340]

On the regulatory front, the USG is acknowledging the calls for greater communication and transparency in its various industry-facing processes. The Department of Commerce, for example, is attempting to provide increased transparency when it comes to export control licenses and approvals. However, while these efforts are welcomed, several industry stakeholders highlighted that the domestic regulatory framework often serves as a hindrance to innovation, is cost prohibitive, and is generally uncoordinated, under-resourced, and bureaucratic across government.

## KEY ISSUES & CHALLENGES

Across the U.S. government, there is strong recognition[341] that Departments and Agencies need to refresh existing tools, introduce new financial instruments, and design new interventions to support U.S. competitiveness in critical and emerging technology sectors.[342] Critically, these efforts will need to work more closely with venture capitalists, traditional banks, entrepreneurs, and technologists to ensure that the money is seeding a competitive technology industrial base for the future and being paired with the right mix of procurement opportunities and international market access to make sure the investments are enduring.

**National Investment Policy** - The U.S. currently does not have a coherent investment policy for the nation, particularly when it comes to Space. While OSTP in the Executive office of the President serves the critical role of coordinating and providing overall strategic guidance to federal R&D funds for all S&T agencies, the system of investment in the U.S. is highly federated. Treasury, OMB, Commerce, NSF, DoD, DoE each make financial injections into the market, but without any coordination or without a roadmap.

---

[338] EXIM (2022). Make More in America Initiative.
[339] AFRL (2021). SpaceWERX launch drives AFWERX small business focus on universities and on-orbit capability.
[340] Magnuson, S. (2022). SOFIC NEWS: Shyu Announces New Fund to Help Small Businesses. National Defense Magazine.
[341] White House (2022). Statement by National Security Advisor Jake Sullivan and National Economic Council Director Brian Deese on EXIM Bank's Domestic Finance Initiative.
[342] Department of Commerce (2021). Department of Commerce Accomplishments Space and Space Commerce. Office of Space Commerce.

**Lack of Patient Capital** - Stakeholders from across the U.S. commercial and national security space ecosystem identify the lack of patient capital in the private sector as a significant bottleneck to developing, maturing and commercializing dual-use space assets. The U.S. government historically offered several tools to de-risk and encourage investments in longer-horizon technologies. Dating back to the Cold War era, the U.S. government certified Small Business Investment Companies (SBICs) which catalyzed investment into the first semiconductor companies in Silicon Valley. Tools such as the Small Business Innovation Research (SBIR) Grants were introduced in the early 1980s and identified promising research, providing seed funding for development of notable American technology companies such as Qualcomm and Symantec. The Department of Energy leveraged its loan authorities to sustain Tesla, Inc., but was also heavily scrutinized for the loss of tax-payer dollars through the failure of solar cell-maker, Solyndra.[343] The Defense Department employs specific tools such as multi-year procurement contracts and capital injections for specific military-exclusive technologies through its DPA Title III authorities.[344]

Figure 59. Axiom Station, the world's first commercial space station, is designed as the foundational infrastructure enabling a diverse economy in orbit. (Credit: Axiom Space)

**Space Infrastructure** - The U.S. government should support the development of space infrastructure by deeming it critical infrastructure. Whether through direct investments or technical resources, the industrial base requires large scale infrastructure investment to develop here on Earth and in space.

**Streamlined and Transparent Regulatory Environment** - The domestic regulatory framework with respect to export controls and licensing often serve as a hindrance to innovation, are cost prohibitive, and are generally uncoordinated across government. The regime could generally be characterized as lacking agility and under-resourced. While there are several overlapping policy priorities that these

---

[343] DoE (2022). Loan Programs Office | Department of Energy
[344] Assistant Secretary of Defense (2022). Assessments & Investments Defense Production Act (DPA) Title III. BuisnessDefense.gov.

regimes are designed to address, they were designed in a previous era when technology was slower to develop, disseminate, and be exploited. Today, commercial innovation cycles are rapidly overtaking regulatory oversight timelines and the various regimes need to be re-thought or considerably reformed to match the pace of innovation.

**Space Activity Treated as a Priority Export Activity** - The ability of the commercial space industry to rapidly strengthen its global competitiveness is in part dependent on the measures the U.S. government puts in place to safeguard its national security. Previously, the U.S. space industry, particularly the satellite industry, has been vocal about the negative impact of export controls on its global expansion, leading to subsequent reforms to export controls. U.S. economic and financial agencies should make space activity a priority export market. This should include trade agreements, trade missions, diplomatic, economic and financial support to U.S. companies seeking to export to international markets.

## KEY INFLECTION POINTS

**U.S. pursues regulatory reforms that align with the pace of technological innovation** and diffusion to allow more coordinated and streamlined decision making processes domestically and internationally related to export controls and investment screening.

**U.S. maintains status quo regulatory policies domestically and internationally**, stifling innovation and enabling foreign competitors who are not encumbered by ITAR or other restrictions from gaining market share.

**U.S. Congress passes Bipartisan Innovation Act** which includes multi-year funding for R&D and commercialization of critical and emerging technologies such as microelectronics, Artificial Intelligence, autonomous systems, and robotics that will underpin the future competitiveness of the U.S. space industry.

**U.S. Government investment capital and debt interventions remain uncoordinated** and disbursed in small dollar amounts, unable to shift the risk calculus for deep-tech space investments and enable more coordinated nation-states to overtake the U.S. to become the global leader in space.

**U.S. Government investment capital and loan facilities are coordinated and concentrated** against high-capital expenditure space investments, significantly reducing risk and crowding in private investment.

**U.S. Government deems Space infrastructure to be Critical Infrastructure**, ensuring that federal technical, financial, and security resources are dedicated to its development and maintenance irrespective of Administration or political winds.

Figure 60. Forthcoming Civilian Missions to the Moon. (Source: NASA / Dr. Rob Landis)

## Fast Follower Strategy for Winning the New Space Race

*"I keep this chart handy as a reminder why we need cislunar space domain awareness today, and why the USSF needs to 'keep up' with the technology, innovation, standards, policy, norms, etc. that this human expansion to the Moon and beyond represents."* - DR. JOEL MOZER, USSF

The Department of Defense has a long history of tackling challenges at the strategic, operational and tactical level by developing new technologies as a 'first mover', with the military as the intended customer and development partner. Examples of this abound and normally dominate the news media (i.e. stealth fighters, missile defense systems, etc.). Such solutions are driven by bespoke requirements, acquired under the Federal Acquisition Rules, and programmed over years as a Program of Record. Alternatively, many new and emergent technology solutions can be sourced from the commercial marketplace. Rapidly adopting and fielding existing commercial technologies is a 'fast follower' strategy[345] that requires no formal definition of requirements, and can be procured and fielded within 2 years under an Other Transaction Agreement using a more flexible and responsive budget process. In other words, the fast follower rapidly accesses technology which the government did not invent.

What enables a fast follower strategy for new and emergent space capabilities? First, a dedicated organizational home for commercial space technology. This enables a consistent ability to assess and procure commercial solutions from a large pool of qualified vendors with frequent refreshes. Procurement of commercial services for launch, remote sensing, communications or domain awareness can all be accomplished without the design, build and sustainment costs that typify large government acquisition programs of record. Instead, service acquisition offices would budget for Capabilities of Record that adapt commercial solutions to emerging threats, enabling agile technology refreshes in line with commercial production cycles. It's been six decades since the establishment of the defense planning, programming and budgeting system (PPBS) process. DoD can modernize by adopting more commercial technology as a fast follower.

---

[345] Brown, M. (2022). Keynote presentation: Fast Follower Strategy: Adopting Commercial Technologies within DoD. SSIB'22 Workshop.

## KEY ACTIONS & RECOMMENDATIONS

These recommendations reflect the numerous inputs from workshop participants and their best assessment regarding which agency(cies) are in the best position to forward the necessary change.

### SHORT-TERM PAYOFF

**Develop a Commercial Space Roadmap that Executes a National Space Investment Policy** - Create a central office or body that would have a "policy planning" type mandate to develop the commercial space roadmap. This roadmap which would, among other things: identify the technologies requiring funding; identify the resources required to support future industries; Identify the offices across government that would be required to take action in a coordinated manner to disburse the resources and enable the industries. This office should be at the Office of Space Commerce at the Department of Commerce; this office would need to be elevated out of NOAA and become a direct report to the Secretary, be provided with greater resources, and a mandate to direct work across the government. (OPRs: National Space Council, DoC)

**Elevate the Office of Space Commerce** - Elevate the Office of Space Commerce outside of NOAA to a role of higher influence and responsibility and empower this agency to develop a commercial space roadmap that would execute a nation space investment policy. This office could also be the central space licensing agency. (OPRs: DoC, National Space Council)

Figure 61. First inaugural meeting of the National Space Council under the Biden Administration in December 2021. (Credit: NASA)

**Optimize SBIC programs to target capital-intensive space technologies** - The USG is currently evaluating how its investment tools can generate greater impact in domestic economic and national security priority areas. The SBIC program previously played an outsized role in jumpstarting the US semiconductor industry, but has waned in influence in the technology markets with the rise of private equity sources. As private investors have largely shifted to software-based technologies, the SBIC program can have renewed importance in encouraging private investors to support "deep-tech" companies, many of which largely reside in the space sector. (OPRs: SBA, DoD/SBIR, OMB, DoC)

**Increase Transparency and Communication in Regulatory Application Process** - Many companies find the U.S. regulatory system cumbersome, slow, and opaque. Some USG agencies have implemented metrics for approvals and provide tracking status via a portal for regulatory and license review and approvals. However, if the approval deviates from the standard process, which they often do, the portal and existing metrics are no longer used. This leads to frustration in the industry and lack of confidence in the process. At this point, transparency is lost between the government and industry. The lack of transparency can further delay approvals, as simple questions or clarifications to the requestors may be unresolved for days. (OPRs: DoC/BIS, DoD/DTSA, DoS/DDTC, DoT/FAA, FCC)

**Constantly review ITAR lists and Remove Technologies Available in the International Marketplace and Treat Space as a Priority Export Activity** - The list of restricted exports needs to be constantly reviewed and updated on a one-to-three-year cadence to ensure that the technology does require controls. If a company believes their technology is dual use, or commercially available, a two-page white paper should be adequate to force an expedient export review. (OPRs: DoC/BIS, DoD/DTSA, DoS/DDTC)

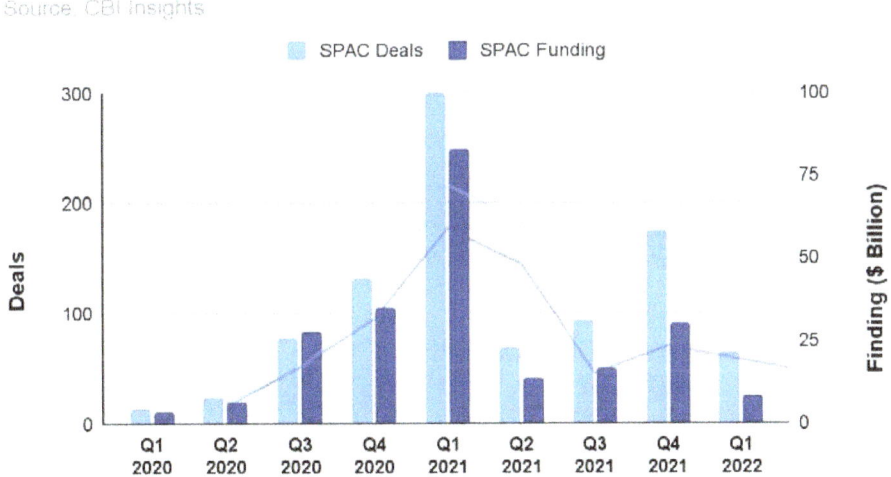

Figure 62. The rise and fall of Special Purpose Acquisition Company (SPAC) transactions. (Credit: DIU)[346]

## MID-TERM PAYOFF

**Inventory and Coordinate USG Investment Tools and Interventions** - Identify the various sources of non-dilutive capital, procurement, advance-market commitments, and debt instruments the USG can provide and align them to the taxonomy of challenges different elements of the space-industry encounter. Make concentrated interventions in these sectors, designed to solve the unique nature of the

---

[346] CBI Insights (2022). What is a SPAC?.

funding challenges. (OPRs: SBA, DoD/NSIC, DoD/DPA Title III, DoD/SBIR/STTR, DoE, ExIm, DFC)

**Develop and Apply Economic Models and Financial Engineering to Incubate and Mature the Space Economy** - Conceptualize new economic models and future technologies that will require financial engineering and financial resources to incubate and mature. (OPRs: National Space Council, OSTP, DoC, DoD/A&S, DoD/USSF, U.S. Ex-Im Bank, Treasury, SBA, NSF)

**Establish Single Licensing Agency or Single Face to the Customer** - The U.S. regulatory process is exceedingly slow and opaque, stifling innovation domestically and between the US and international partners. In general, there is currently not a single point of approval or denial, and it is often difficult to achieve consensus between the necessary regulatory agencies (e.g., FAA, FCC, Dept of State, Dept of Commerce). Any agency can delay or deny approvals. This process lacks transparency and takes a 'red light' approach to approvals. Establishing a single licensing agency will enable a "single point of contact" for industry, concentrate government regulatory expertise, and accelerate the process and transparency. (OPRs: DoC, FAA, FCC)

**USG needs to identify space infrastructure as critical infrastructure** – a recommendation that carries over from previous SSIB reports. Space infrastructure should be considered critical infrastructure per the USG. The USG should lead investment in space and technology infrastructure to drive the future economy and growth. Companies locate in areas where infrastructure exists, for example, semiconductor facilities. (OPRs: DHS, National Space Council, NSC, NEC, DPC)

**Space Stakeholders Must Engage and Shape Federal Resources Disbursed for Infrastructure and Technological Competitiveness** - Various Departments and Agencies across the US Government will be charged with executing the $1+ trillion in federal funding for infrastructure and the proposed $200+ billion in funding for critical and emerging technologies. Stakeholders across the space ecosystem have an opportunity to shape how these monies are spent and should make every effort to ensure that relevant space infrastructure is prioritized for federal funds and technical resources. (OPRs: National Security, Civil, Commercial Space Stakeholders)

**Supply Chain Development and Understanding** - Consistent with the Administration's focus on building more resilient supply chains in microelectronics, pharmaceuticals and advanced batteries, the space industry needs greater clarity on the health, strengths and vulnerabilities of critical components in its supply chain. The supply chain for satellites, launch infrastructure, advanced communications and other critical space-enabling technologies should also be considered critical as part of space infrastructure. It is critical that supply chains are understood, supported, and trusted between public and private partnerships. (OPRs: DoC, DoD, DoE)

**Target International Engagement Around Specific Topic Areas: U.S. international engagement in space remains vague** - The U.S. should specify areas for collaboration, develop roadmaps and measure progress. For example, the U.S. could identify space traffic management or space situational awareness as specific topic areas, where each partner can focus technical, financial, and diplomatic resources to improve collective capability and interoperability. (OPRs: OSTP, DoS, DoC/Office of Space Commerce, DoD/Office of Space Policy, USSF)

LONG-TERM PAYOFF

**Simplify DFARs to Lower Barriers to Entry and Accelerate Acquisitions to Speed of Relevance**
- The Defense Federal Acquisition Regulation Supplement (DFARS) to the Federal Acquisition Regulation was written in the 20th century and has become more complex, burdensome over time. The DFARS requires a review to determine the intent of the requirement, what risk does removing the regulation pose, and is the requirement applicable. Ideally, this review would reduce and simplify the DFARS significantly, making it easier for commercial companies to do business with the USG/DoD. (OPR: Congress, DoD)

**USG should designate space as an economic opportunity zone to encourage growth and investment** - The U.S. can designate Space as an economic opportunity zone in order to improve the economic prospects for critical space-technologies or infrastructure. Historically, U.S. economic zones offered a range of benefits, including favorable tax treatment for investors and companies, improved trade opportunities, or the ability to leverage federal infrastructure. In order to advance U.S. space competitiveness, policy makers need to consider such a designation in order to incent stakeholders across the space community to actively invest in space. (OPRs: Treasury, OMB, Commerce, Congress)

**Follow the airport model for space infrastructure** - Similar to how the federal government defers the planning and permitting of airports to the local level, the USG should work with state and local officials to ensure that states/regions determine the best use of funds, with light-touch direction from the federal government. The local regions are responsible for future planning and growth and are therefore the appropriate agency to plan space infrastructure. (OPRs: DoT, DoC, State and Local Governments)

Figure 63. California Governor Gavin Newsom touring Astra Space, Inc. (Astra) to highlight the economic and technological advancements California is supporting in partnership with space companies. (Photo courtesy of Astra and the Office of Governor Gavin Newsom)

Figure 64. Six unidentified NASA scientists use ladders and a large chalkboard to list reference equations at Systems Labs, California, 1957. (Credit: J. R. Eyerman, LIFE)

# WORKFORCE & STEM EDUCATION

*"The future of the economy is in STEM,
that is where the jobs of tomorrow will be."*

— JAMES BROWN[347]
Director, STEM Education Coalition

## BACKGROUND

Given today's imperatives to grow the space industrial base for national security, commercialization and sustainability, a more robust workforce must be put in place for the U.S. to win the second space race. China has made space a key element of their national strategy and their progress is evident in the increasing strength of their science, technology, engineering and math (STEM), or STEAM with the arts included, workforce – which outpaces all nations in growth and size. China sits at approximately 140% of the total number of scientists and engineers (S&Es) compared to the U.S., and that number becomes larger every year. Due to vast differences in the number of S&Es and the sheer size of the population between the two countries, the U.S. needs to focus on attracting and retaining larger numbers towards STEM fields, with an emphasis on directly supporting space exploration, expansion and defense. While the focus on workforce development in the STEM fields is the immediate and most critical goal, the U.S. should also eye development in other non-technical areas which support space-STEM, such as business, finance and the arts (STEAM+).[348]

**Leading and winning the new space race will require increasing the size of the workforce** through greater access to education and reaching untapped, underrepresented populations. The U.S. must also organize and prioritize focus areas for space innovation and commercialization, and align workforce pipeline development accordingly given the advantage China has by way of sheer numbers in R&D.

**China is Growing Researchers in R&D Twice as Fast as the U.S.** - The U.S. and China are now, approximately, both graduating and employing the same number of researchers per thousand in population. This puts the U.S. at a tremendous disadvantage as the population of China is over three times as large.

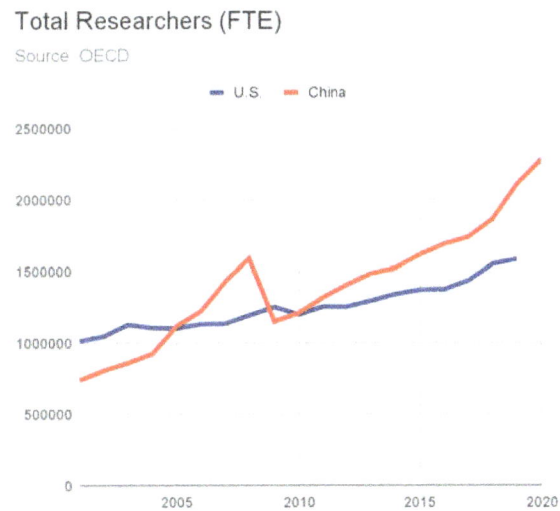

Figure 65. China and the U.S. graduate and employ the same number of full-time equivalent (FTE) researchers per thousand in population, giving China a threefold advantage due to its exponentially larger population size.[349]

---

[347] Vilorio, D. (2014). STEM 101: Intro to tomorrow's jobs. U.S. Bureau of Labor Statistics.
[348] Marr, B. (2020). We Need STEAM, Not STEM Education, To Prepare Our Kids For The 4th Industrial Revolution. Forbes.
[349] World Bank (2022). Researchers in R&D (per million people).

As a result, China adds considerably more researchers to their workforce per year compared to the United States. China surpassed the U.S. in terms of the raw (absolute) number of people in R&D a decade ago with no signs of slowing down.[350] For decades, the U.S. has led the world in the number of S&E doctorates awarded (41,000 in 2018); however, China is about to close that gap.[351] Due to the threefold population advantage China has, the gap in the number of U.S. science and technology research and development professionals will continue to become worse every year without focused attention, increased financial investment and a long term strategy to reverse these trends.

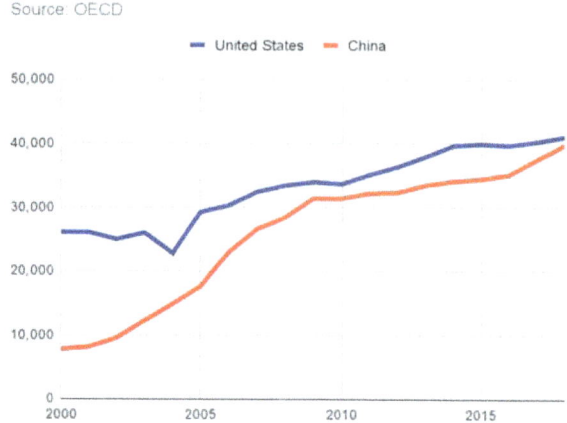

Figure 66. For decades, the U.S. has led the world in the number of S&E doctorates awarded (41,000 in 2018); however, China is closing the gap. (Credit: DIU)[347]

**China Graduates are projected to outnumber USA STEM Grads 15:1 by 2030** - Figures from the World Economic Forum report indicate that China had 4.7 million STEM graduates in 2016 and their graduates outpaced U.S. STEM graduates 8.2 to 1. It is noteworthy that China's investment in STEM also included spending on physical infrastructure at a rate of almost one university per week during this same timeframe.[352] In addition, modest predictions see the wide STEM-graduate gap becoming wider. "By 2030, China and India could account for more than 60% of the STEM graduates in major economies, compared with only 8% in Europe and 4% in the United States."[353]

> *"The United States needs to reinvigorate its STEM education system if it is to compete successfully in the 21st century. STEM proficiency has been declining in America since the 1980s, threatening the nation's continued technological leadership."*
>
> – Center for Strategic and International Studies (CSIS), 2022[354]

## CURRENT STATE

The SSIB'21 report outlined that a healthy space infrastructure supported by a STEM workforce must be put in place to capture a dominant share of the estimated $1.4 trillion in space global economic growth expected over the next decade. Success will require a highly educated workforce across all STEM disciplines. However, workforce issues threaten the economic viability of space as well as the ability to maintain a strong national security space posture. The Workforce, Education and STEM Working

---

[350] World Bank (2022). Researchers in R&D (per million people).
[351] NSF (2022). The State of U.S. Science and Engineering 2022. The National Science Board.
[352] World Bank, Researchers in R&D (per million people).
[353] Wang, B. (2017). Future Tech dominance – China outnumber USA STEM Grads 8 to 1 and by 2030 15 to 1. Next Big Future.
[354] G. Athanasia and J. Cota (2022). The U.S. Should Strengthen STEM Education to Remain Globally Competitive. CSIS.

Group was added as a new working group in 2022 to address issues identified in previous SSIB reports regarding how the lack of a space-ready workforce would substantially inhibit progress. While much work needs to be accomplished to address the complexity required in building a larger space-STEM and STEAM+ workforce pipeline, momentum to progress STEM workforce development has increased in recent years.

**The Administration Makes Historic Investments in STEM** - The Administration announced increased and historic spending in STEM for the last two years. Most recently, the FY 2023 budget has spending levels for basic and applied research that top $100 billion and spending for total federal R&D that tops $200 billion.[355] The budget includes $150 million for NASA's Office of STEM Engagement to attract diverse groups of students to STEM through learning opportunities that provide connections to NASA's mission and work.[356] These investments come after recent announcements for $39 billion in funding through the STEM Education Boost in American Families Plan and historic investments in STEM in FY 2022.[357]

**The National Space Council Releases Space Priorities Framework** - The first National Space Council meeting chaired by Vice President Harris was held in December 2021, with the U.S. Space Priorities Framework released afterward.[358, 359] The National Space Council pledged as a critical priority to continue U.S. leadership into the next generation and prepare the nation's STEM workforce to fuel the economy of the future.[7] Under this umbrella, sub-priorities will be needed to build the STEM talent pool most effectively. One workgroup participant commented, "we can't sustain being number one in everything when we are outnumbered in STEM."

> *"Across the U.S., more than half of high schools do not offer calculus, four out of ten do not offer physics, more than one in four do not offer chemistry, and more than one in five do not offer Algebra II (a gateway class for STEM success in college)."*
> – Center for Strategic and International Studies (CSIS), 2022[360]

**Efforts to Attract Skilled Workforce are Fragmented and Complex** - The working group acknowledged the complexity and fragmented nature of the U.S. STEM ecosystem, making it difficult to achieve a unity within academia, industry and government. As a result, workforce needs from industry go unmet for multiple reasons, including a lack of graduates and critical skill sets, retention issues and government clearance requirements. Industry has trouble finding space professionals, particularly those that have reached mid-career. New graduates and near-retirement professionals are more prevalent but have shortfalls in experience or are looking to retire in the near future respectively. One barrier identified by industry and academia is that government contracts for classified projects require that all full-time employees be cleared at the contract level, even if portions of the work are unclassified. This prevents participation by uncleared interns, students, and junior professionals. In

---

[355] White House (2022). The Biden-Harris Administration FY 2023 Budget Makes Historic Investments in Science and Technology.
[356] Smith, J. (2021). Biden Aims $39 Billion at STEM Education Boost in American Families Plan. MeriTalk.
[357] Dennon, A. (2021). Biden's Edu Budget Prioritizes STEM, Students of Color. Best Colleges.
[358] State Department (2021). Vice President Harris' First National Space Council Meeting.
[359] White House (2021). United States Space Priorities Framework.
[360] G. Athanasia and J. Cota (2022). The U.S. Should Strengthen STEM Education to Remain Globally Competitive. CSIS.

addition, the scarcity of access to mid-level space professionals to mentor junior-level talent makes it harder for junior professionals to progress their careers and contributes to unmet employer needs. Onboarding skilled workers from outside the U.S. also is a major challenge for government space projects, specifically in the DoD, due to over-classification and restrictions from International Traffic in Arms Regulations (ITAR). Classifying authorities within the government appear to be too risk-averse or overstretched when it comes to downgrading or declassifying space projects/programs.

> *"The U.S. is at risk of losing $454 billion of manufacturing GDP in 2028 alone due to a skills shortage."*
> – DELOITTE and the MANUFACTURING INSTITUTE, 2018[361]

**Job Positions Go Unfulfilled, Yet Diverse Talent Pools Are Untapped** - Academia, industry, and government within the space sector struggle to fill space-STEM positions while opportunities to overcome this challenge are overlooked. For instance, the U.S. could greatly improve the size of its talent pool with successful outreach to underrepresented populations. Another observation given from a workshop participant described how there is a shortage of creativity or willingness from academia, industry, and government to investigate other sources of technical talent from adjacent fields. Despite having a technical foundation, professionals in declining industries that would need mild to moderate reskilling for space-STEM are overlooked. In addition, many non-STEM jobs need to be filled, such as contract managers and financial engineers, but they remain glossed over in talent pipeline development efforts because filling STEM jobs takes precedence. As a result, today's workforce grows at a much slower rate than possible, and the lack of diversity in the talent pipeline that would otherwise be possible through effective STEM education. The industry is starting to unite on a greater scale to better address these issues. Space Workforce 2030 launched at the Space Symposium in April of this year, where over 25 prominent space companies pledged to advance diversity, equity and inclusion in the space industry.[362]

> *"I was hesitant to sign that pledge because I wanted to see what actual work was going on, because there's a lot of virtue signaling in a lot of these things. I'm the father of an eight-year-old daughter and I see the stereotypings literally evolving in front of my eyes. So at Rocket Lab many years ago we said, 'Right, we're going to mandate that 50% of all our interns are going to be female.' And that's great, you can mandate something, but if you don't do the work it's a complete waste of time. So the team went out and we visited, I think, something like 200 schools… you have to get in super early, you've got to build the pipeline. Actually create the change, not just sort of sign onto change."*
> – PETER BECK, CEO and Founder of Rocket Lab, 2022[363]

---

[361] Deloitte (2018). 2018 Deloitte and The Manufacturing Institute skills gap and future of work study. Deloitte.
[362] Space Workforce 2030 (2021). Space Workforce 2030. SWF2030.
[363] Coldewey, D. (2022). Major space companies pledge to boost diversity and publicly share hard numbers. TechCrunch.

Figure 67. Mentorship counts. STEM students are inspired to succeed with heroes (Photo: NASA)

## A rigorous, six-week, in-resident STEM program with Mentors

Reading this section on workforce and the current state of STEM education can leave one emotionally exhausted and depressed about the future. As Mark Twain once observed, "the truth hurts, but silence kills." Clearly more can be done to achieve measurable results. A recent paper for the National Bureau of Economic Research found that success in higher education was strengthened by participation in a mentored, summer program focused on STEM.[364] Equally important, such programs are particularly impactful for underrepresented youth. Upon evaluation of a variety of STEM programs (including some offered online), this report lauded a rigorous, six-week, in-resident STEM program sponsored at a university, stating:

> "Students offered seats in the STEM summer programs are more likely to enroll in, persist through, and graduate from college, with gains in institutional quality coming from both the host institution and other elite universities. The programs also increase the likelihood that students graduate with a degree in a STEM field, with the most intensive program increasing four-year graduation with a STEM degree attainment by 33 percent." - SARAH R. COHODES, HELEN HO, and SILVIA C. ROBLES

Why? In part, because the most successful students bond with a mentor during the in-resident summer program taken between their junior and senior years of high school. It also led to increases in enrollments at the nation's most competitive colleges. Eighty-seven percent of the participants attended a four-year college immediately after high school graduation; and 34% graduated with a STEM degree within four years.

Notably, the program achieves success by increasing access to STEM careers, exposing students to high-achieving peers, STEM mentors, STEM curriculum, tours of a college campus and research facilities, and college admissions information. It immerses rising high school seniors in rigorous science and engineering classes including courses in math, physics, life sciences, and humanities, as well as a STEM-related elective course with topics ranging from digital design to genomics. In addition, students take tours of labs and work spaces at an elite university; attend workshops with leaders of industry and academics and admissions officers; and interact with teaching assistants who are current college students. Students also visit STEM-focused companies and workplaces. Our nation needs more programs and mentors like these. Bravo!

---

[364] Cohodes, S. et al (2022). STEM summer programs for underrepresented youth increase STEM degrees. NBER.

## KEY ISSUES & CHALLENGES

The most significant barriers to attracting and retaining the space workforce include lack of awareness of the full range of job opportunities, equitable access to education and jobs, stringent hiring requirements, funding gaps and generational issues.[365] In addition, the fragmented nature of the space-STEAM and STEAM+ ecosystems adds complexity to these issues and creates obstacles in progressing a larger pool of talent – from early education, through higher education, to early and mid-career. Among the many nation-wide STEM and STEAM+ programs, only a paucity are specifically space-focused, and generally these lack the scale, reach, penetration into underrepresented communities to sustain and grow the talent pipeline. Examples of such programs include SpaceCamp,[366] Federation of Galaxy Explorers,[367] Space Art Competitions,[368] and the NSS[369] and International Space Settlement Design Competition.[370] Neither NASA nor the USSF has the equivalent of a Civil Air Patrol cadet program[371] for youth. No colleges have dedicated programs or endowed chairs for space industrialization.[372] Few community colleges focus on building skills specific to the space community.[373]

---

*"Many, if not most, of the foreign nationals earning advanced STEM degrees from U.S. universities would prefer to stay and work in the country. But America's immigration system is turning away these workers in record numbers — and at the worst possible time."*

– POLITICO, 2022[374]

---

**Attracting Workforce is Complex & Incomplete** - STEM research indicates that children should begin learning about the world of STEM by the time they are in fifth grade or it may be too late to have them become interested in a STEM career.[375] However, outreach to younger students requires more human and financial capital and has special considerations. Efforts to inspire children under nine years

---

[365] Beyer, B. and Dittmar, M.L. (2018). Op-ed: Talent gap jeopardizes space business, national security. SpaceNews.
[366] U.S. Space & Rocket Center (2022). SpaceCamp.
[367] Federation of Galaxy Explorers (2022). Federation of Galaxy Explorers.
[368] Space Foundation (2022). Space Foundation Art Contest.
[369] NSS (2022). NSS Space Settlement Contest.
[370] SpaceSet (2022). Space Settlement Design Competition.
[371] Civil Air Patrol (2022). Cadet Programs | Civil Air Patrol National Headquarters.
[372] NASA previously sponsored a University Space Engineering Research Centers (USERC) program. The purpose of the USERC program was to replenish and enhance the capabilities of our Nation's engineering community to meet its future space technology needs. The Centers were designed to advance the state-of-the-art in key space-related engineering disciplines and to promote and support engineering education for the next generation of engineers for the national space program and related commercial space endeavors. Research on the following areas was initiated: liquid, solid, and hybrid chemical propulsion, nuclear propulsion, electrical propulsion, and advanced propulsion concepts. The USERC program was initiated in 1988 by the Office of Aeronautics and Space Technology to provide an invigorating force to drive technology advancements in the U.S. space industry and operated till at least 1995 at about $13.6M/yr. In the 1995 NASA Authorization Act, Congress stated: "The committee deplores the fact that NASA plans to zero the funding for the University Space Engineering Research Centers (USERC's) after fiscal year 1995."
[373] One example is the NASA Marshall Space Flight Center Partnership with Drake State Community college to advance their Moon-to-Mars Planetary Autonomous Construction Technology (MMPACT) project; Drake State (2020). NASA awards Cooperative Agreement to Drake State Community & Technical College for Space Construction Research; Drake State (2021). Drake State awarded nearly $1.2 million by NASA.
[374] Bordelon, B. & Mueller, E. (2022). Biden wants an industrial renaissance. He can't do it without immigration reform. Politico.
[375] Purdue University (2019). Early Exposure to STEM and its Impact on the Future of Work.

old may not be well received by parents who view it as forced learning. In addition, young people may not resonate with a message from an outreach professional who is significantly older than them or does not look like them. The lack of diversity in today's workforce is slowing down the growth of the future workforce. Alongside this challenge, there is a lack of awareness across large swaths of the U.S. population regarding the full range of possibilities and opportunities for a space career.

> *"Although Blacks or African Americans, Hispanics or Latinos, and American Indians or Alaska Natives represent 30% of the employed U.S. population, they are 23% of the STEM workforce due to underrepresentation of these groups among STEM workers with a bachelor's degree or higher."*
>
> – NATIONAL SCIENCE FOUNDATION, 2021[376]

**Retention Issues Stifle Workforce Growth** - Messaging and outreach issues also impact retention. Workers decide to leave the industry due to a lack of awareness and access to continuing education, the allure of competing industries, and workers that do not see others like themselves inside a company, in outreach positions, or in the industry at large. Some companies are taking steps to address diversity in their organizations. However, if staff or job candidates see a diversity-focused recruiting effort as inauthentic or pandering, the result contradicts its intention with alienated individuals or communities.

**Access to Education Isn't Equitable** - More investment is required to attract a larger, more diverse workforce with better representation from minority and underserved populations. Two key challenges identified by participants that prevent access to education were funding support to low-income families and more 'wrap-around' services from colleges to support students who may drop out (e.g., parents lacking affordable child care, students lacking reliable transportation).

**Substantial Financial Gaps Persist** - While recent Administrations have increased STEM spending to historic levels[377] – including more support for R&D, physical infrastructure, women and minorities – financial gaps persist on the ground for developing the future workforce.[378] Participants noted several gaps that prevent space workforce growth, including the need for better physical infrastructure, grant programs for technical trade programs, stronger communication of financial support resources, cultural sensitivity training, and not enough paid internships. U.S. infrastructure is both overstretched and lagging behind that of its peer nations and competitors, particularly China.[379]

**Younger Generations Have Different Expectations** - Younger generations (millennials and Gen Z) have different values than preceding generations, shaped in part by unique historical events that have happened in their lifetimes. Academia, industry and government must adjust their recruitment and retention strategies accordingly so they do not falter. Expectations from younger generations include remote work where possible, work-life balance, paid time-off, increased parental leave, access to childcare, and pay equity and transparency within companies.

---

[376] Okrent, A. & Burke, A. (2021). The STEM Labor Force of Today: Scientists, Engineers, and Skilled Technical Workers. NSF.
[377] U.S. Department of Education (2022). Science, Technology, Engineering, and Math, including Computer Science.
[378] Smith, J. (2021). Biden Aims $39 Billion at STEM Education Boost in American Families Plan. MeriTalk.
[379] Clifford, M., Kwok, D., Lim, L., & Brouillette, D. R. (2021). The State of US Infrastructure. CFR.

## KEY INFLECTION POINTS

**The U.S. fails to reverse the increasing STEM-Graduate gap with China** - Due to workforce pipeline issues, China pulls further ahead in research-and-development investment, manufacturing critical emerging technologies, patents for innovative systems, and conferring bachelors up through PhDs in STEM.

**The U.S. succeeds in exponentially growing the number of R&D researchers** - The Administration's increased STEM R&D budget successfully flows to the space sector to reverse current R&D workforce trends.

**The U.S. provides financial backing to right-size space-STEM investment** - Historic budget investments enable federal agencies and the National Space Council to right-size investments to address funding gaps across physical infrastructure, equitable access to education, and reaching and attracting a larger and more diverse talent pool for the space sector.

**The U.S. succeeds in removing domestic and international recruitment barriers** - The U.S. opens up a larger talent pool and closes the gap with Chinese STEM matriculation rates by addressing over-classification and restrictions of domestic and international students that prevent hiring or career advancement.

**The National Space Council launches a North Star Vision with Public/Private partnership to grow the space workforce** - The vision bolsters progress in achieving the goals of the National Science & Technology Council's STEM Education Strategic Plan, unites fragmented workforce efforts in alignment with employer needs and removes barriers to entering and progressing in STEM fields.

---

> *"One thing business leaders and educators readily agree on is that if we are to have sustained growth in the space industry, we must have an uninterrupted pipeline of talent… The jobs available in the global space ecosystem are becoming more varied and increasingly technical in nature and are destined to help create new products and services both in space and on Earth. If we are to realize that growing potential, we must have the talent pool to get us there."*
>
> – THOMAS "TOM" ZELIBOR (Rear Admiral, USN, ret), CEO, SPACE FOUNDATION, 2022[380]

---

## KEY ACTIONS & RECOMMENDATIONS

### SHORT-TERM PAYOFF

**Pilot Pathways to The Stars** – Create a space workforce pathway spanning early childhood to early career with industry engagement, education provider partnerships, and an emphasis on outreach and support that is more inclusive of diverse populations. Once in place, the intent is to enable other regions

---

[380] Space Foundation (2022). The Space Report 2022 Q1. Space Foundation.

to replicate and tailor the program for their regions. (OPRs: NewSpace New Mexico (Space Industry), SSIB Workforce Education and STEM Working Group, Space Workforce 2030)

**Create strategic messaging at a national level** that can be used by all regions and sectors of the industry to showcase the full range of job opportunities in space, and past successes from minority individuals in space careers, to attract and retain a larger, more diverse workforce. Alongside this process, harness the "wow" factor of space in messaging efforts to increase returns. (OPRs: NewSpace New Mexico (Space Industry), SSIB Workforce Education and STEM Working Group)

## MID-TERM PAYOFF

**Bolster the U.S. STEM Strategic Plan with an Aligned Space-STEM North Star Vision**[381] - Addressing the complexity and fragmentation of space-STEM and STEAM+ educational systems requires "centralized planning, decentralized execution." Fixing, expanding and extending workforce development to fill identified gaps will demand agreed-upon priorities and alignment across academia, industry and government, led by the National Space Council with public/private partnership. Whole-of-nation STEM workforce development strategies can be executed at a local level to match the needs of the thousands of diverse communities around the country in order to maximize recruitment and retention in space-STEM fields. (OPRs: National Space Council, National Science and Technology Council Committee on STEM Education, SSIB Workforce Education and STEM Working Group, DoD, NASA-STEM Engagement, NewSpace New Mexico (Space Industry))

**Assess and Address Funding Gaps** - to match the desired outcomes in increasing the size of the space workforce in alignment with national security, commercialization and sustainability imperatives. (OPRs: Office of Personnel Management (OPM), DoD, NASA, Industry, National Science and Technology Council)

## LONG-TERM PAYOFF

**Scale Pathways to the Stars Program** – With whole-of-nation STEM workforce development strategies in place, the Pathways to the Stars program can be replicated in different regions and scale successes from the pilot program. (OPRs: NewSpace New Mexico (Space Industry), SSIB Workforce Education and STEM Working Group, Space Workforce 2030)

---

[381] Committee on STEM Education, National Science & Technology Council. (2018). Charting a Course for Success: America's Strategy for STEM Education.

*"Today, at laboratories, on launchpads, and in orbit — often, in partnership with our government — commercial space companies are making real the opportunity of space for millions of Americans. Their work is accelerating innovation in the space sector and shaping our nation's future in space."*

– VICE PRESIDENT KAMALA HARRIS, 2022[382]

Figure 68. Vice President Harris addresses members of the commercial space industry at the Chabot Space Science Center in Oakland, CA. Accompanying her (left to right) are Governor Gavin Newsom, FAA Acting Administrator Billy Nolen, and Federal Communications Commission Chairwoman Jessica Rosenworcel. (Credit: Dominic Hart, NASA)

---

[382] White House (2022). Remarks by Vice President Harris On Supporting the Commercial Space Sector. The White House.

# EPILOGUE

Figure 69. James Webb Telescope Image of the Carina Nebula (Credit: NASA)[383]

---

*"This day I completed my thirty first year, and conceived that I had in all human probability now existed about half the period which I am to remain in this Sublunary world. I reflected that I had as yet done but little, very little indeed, to further the happiness of the human race, or to advance the information of the succeeding generation. I viewed with regret the many hours I have spent in indolence, and now soarly feel the want of that information which those hours would have given me had they been judiciously expended. but since they are past and cannot be recalled, I dash from me the gloomy thought and resolved in future, to redouble my exertions and at least endeavor to promote those two primary objects of human existence, by giving them the aid of that portion of talents which nature and fortune have bestowed on me; or in future, to live for mankind, as I have heretofore lived for myself."*

– MERIWETHER LEWIS, 1805[384]

---

## CONCLUDING THOUGHTS

This year NASA thrilled us with first images from the James Webb Space Telescope, stationed a million miles from Earth and looking back billions of years, and inspiring a next generation of STEM students. We should consider what the recommendations in this report make possible for future generations: with lower-cost and larger volume launch, next generation power and propulsion, in-space servicing

---

[383] NASA (2022). First Images from the James Webb Space Telescope.
[384] Meriwether Lewis's journal entry for 18 August 1805, while crossing the Rockies.

assembly and manufacture, and in-space transportation, we could build in space far larger telescopes at lower cost. Those telescopes would be serviceable and, like terrestrial telescope facilities, upgradable to retain their technological relevance. For the same price, we could build not one, but many.

**Selfless Aspiration** - Meriwether Lewis did not fully comprehend the impact of his contributions to future generations until he was 31, but his vision was so profound that he dedicated the rest of his life to realizing it. In the spirit of the great American explorers, engineers and entrepreneurs of the past who broke new ground from the trailhead of uncertainty to a more prosperous future that preserves liberty and self-determination amongst a family of nations, now is the time for this generation to 'live for the future of humankind, as we have heretofore lived for ourselves.' There are many distractions that feed division and despair within our nation that only serve to feed the autocratic desires of our strategic competitors. Unity of effort is achieved when Americans appreciate what is at stake, and aspire to apply themselves to charting a better future.

**Per aspera ad astra**[385] - What America needs now is a mission, a purpose that strikes at the selfless soul of both young and old. We must lead. We must prevail. We must trailblaze a path to the Moon, Mars and beyond that creates hope and opportunity for all humankind. Much as America has been the beacon of hope for millions of emigrants throughout our past, the solar system provides new worlds and new opportunities for potentially billions of people in the future. America needs to shape its space future today in order to assure future generations the opportunity to lead from the front in shaping the next great adventure for humanity in the future. A North Star Vision of economic development and human settlement is the first step toward achieving that goal.

> *"We can see possibilities no one has ever seen before. We can go places no one has ever gone before. You know, you've — you've heard me say it over and over again. America is defined by one single word: possibilities. Possibilities."*
>
> – PRESIDENT JOE BIDEN, 2022[386]

---

[385] A latin phrase meaning "through hardships to the stars" or "Our aspirations take us to the stars."
[386] White House (2022). Remarks by President Biden and Vice President Harris in a Briefing to Preview the First Images from the James Webb Space Telescope. The White House.

# APPENDIX A
## WORKSHOP PARTICIPANTS

**LIVE PARTICIPANTS**
Albuquerque, NM

Africano, James; Regolith Ventures
Ahn, Joe; Northrop Grumman
Anderson, Steve; LMI (Logistics Management Institute)
Armendariz, Jorge; New Mexico Spaceport Authority
Armstrong, Jason; TriSept Corporation
Aspiotis, Jason; Axiom Space Inc.
Baker, Austin; DIU
Baker, Julie; Ursa Space Systems Inc.
Banks, Darwyn, NRO
Bargiel, Jeff; CNM Ingenuity | Hyperspace Challenge
Barnaby, Dave; AFRL
Bendle, Klay; DIU
Bever, Marcus; ExoAnalytic Solutions
Blenkush, Severin; Space Advisory Group
Blocker, Jona; USNC-Tech
Brickley, Denise; RESPEC Company LLC
Butow, Steven "Bucky"; DIU
Byers, Wheaton; Verus Research
Caudill, Tom; BlueHalo
Chi, Dan; USAF OCEA
Christodoulou, Christos; University of New Mexico
Cook, T.J.; CNM Ingenuity
Cooley, Tom; AFRL
Cover, Park; Avalanche Energy
Crawford, Meagan; SpaceFund
DeRaad, Casey, NewSpace New Mexico
Dyble, Tom; DYBLE Associates
Enoch, Michael; Lockheed Martin Space
Erwin, Scott (Richard); AFRL/Space Vehicles Directorate
Fischer, Jack; Intuitive Machines
Flake, Richard; Innovative Strategies
Flewelling, Brien; ExoAnalytic Solutions Inc.
Gapp, Nathan; DIU
Garcia, Amanda; Los Alamos National Laboratory
Garretson, Peter; AFRL/RV
Gil, Chris; Northrop Grumman
Glassner, Samantha; DIU
Goswami, Namrata; Independent Scholar
Halbach, Rick; Lockheed Martin
Hannigan, Allie; Xplore Inc.
Hannigan, Russell; Xplore Inc.
Hardy, David; Apogee Engineering LLC
Harris, Robbie; Nanoracks
Hickman, Zachary; Air Force Office of Commercial and Economic Analysis (OCEA)
Hodge, Michael; Lynk Global
Hoffman, Lars; Rocket Lab USA
Hooks, Daniel; Los Alamos National Laboratory
Irby, Rhonda; SpinLaunch
Jacobs, Scott; Astranis Space Technologies Corp.
Jaques, Danny; Danny Rocket Ranch Space Salsa
Johnis, Benjamin; USAF / AFIT
Jones, Johanna; DIU
Joseph, Nikolai, NASA
Kennedy, Bryce; AEGIS Trade Law
Keravala, Jim; OffWorld
Kershaw, Andrew; Lockheed Martin
Killings, M. Leon; Space Systems Command/S3
Kirkendall, Barry; DIU
Kwas, Andrew; Northrop Grumman Corp
Lamanna, Matthew; Deloitte
Lance, Cameo; Rhea Space Activity
Lawless, Juli; Redwire
Luce, Kitty; Northrop Grumman Corp
Martin, Hall; TEN Capital Network
Martin, Tom; Blue Origin
Martinez, Melody; AFRL
Masri-Elyafaoui, Ramzi; Virgin Galactic Holdings Inc.
McAlpine, Bradley; Lockheed Martin Space/Space Security
McClain, Sean; NRO
Metcalf, Andrew; AFRL
Milburn, J. Chris; USSF
Moberly, John; LeoLabs Federal
Mommer, Ric; DIU
Neuman, Jay; Neuvations
Noone, Jordan; Embedded Ventures
Okandan, Murat; mPower Technology, Inc.
Osborne, Kenneth; Space Systems Services, LLC
Pandain, Muk; Varda Space Industries
Papandrew, Devon; STOKE Space
Paul, Chris; AFRL
Peng, Thomas; AFRL/Space Vehicles Directorate
Pereira, Michael; Los Alamos National Laboratory
Perry, Jacob; AFRL
Peterkin, Robert; General Atomics Electromagnetic Systems Group
Poulos, Dennis; DIU

Remen, John; AFRL/RQR
Reynard, Jeremy; DIU
Rich, Jeff; Xplore Inc.
Rich, Lisa; Xplore Inc.
Ridenoure, Rex; DIU
Rideout, Beau; Rhea Space Activity
Riedewald, Jared; SpaceLogistics
Rogers, Lauren; DIU
Salwan, Eric; Firefly Aerospace
Sanchez, Merri; Aerospace Corporation
Sandhoo, GP; DIU
Sanford, Gregory; Redwire
Schaab, Adrienne; AFRL
Schenkkan, Tex; DIU
Schwenke, Jim; SpaceLink
Shidemantle, Ritchard; Hedron
Shimmin, Rogan; DIU
Steen, Kathy; Hyper Space Challenges
Stevenson, Rhonda; Orbital Assembly Corporation
Stolleis, Karl; AFRL
Tanner, Maraia; Star Harbor
Teehan, Russ; AWS
Theret, Tara; Northrop Grumman
Tumlinson, Rick; SpaceFund, New Worlds Institute
Unis, Carl "C.J."; Space Systems Command, USSF
Venneri, Paolo; USNC
Wagner, John; Sierra Nevada Corporation
Wallace, Jason; DIU
Walsh, Steven; UNM School of Anderson
Weed, Ryan; DIU
Wegner, Peter; BlackSky
Winter, James; AFRL/RVEP
Winter, Laura; Defense & Aerospace Report
Wood, Stephen; Maxar Technologies
York, Michael; Microsoft

## LIVE PARTICIPANTS
Cape Canaveral, FL

Armentrout, Taylor; Virgin Galactic
Autry, Greg; Thunderbird/ ASU
Baird, Mark; Virgin Orbit National Systems
Baughman, Adrienne; Sidus Space
Bell, Elizabeth Jordan; GXO, Inc
Bendle, Klay; DIU
Berkson, Brad; RocketStar
Bontrager, Mark; USSF
Buck, David; BRPH Mission Solutions
Carrillo, Ruben; California Military Department / 148th Space Ops Squadron
Chung, Jon; NRO

Daugherty, Brandon; Operator Solutions
Dixon, Josh; Radian Aerospace
Esbeck, Ann; Bechtel National, Inc.
Faler, Wesley; Miles Space, Inc.
Feldman, Raphael; SpinLaunch
Figueroa Rodriguez, Israel; SAIC / NRO/OSL
Fleming, Shane; Rocket Lab
Garcia, Celestino; ENSCO
Golceker, Taylor; Aevum inc
Hook, David; The Aerospace Corp.
Karuntzos, Keith; NRO/OSL
Killings, M. Leon; Space Systems Command
Klinger, Gil; Virginia Commercial Spaceport Authority
Koroshetz, John; Sierra Space
Kulin, Robb; STOKE Space Technologies
Lloyd, Steve; All Points
Marotta, Thomas; The Spaceport Company
Mathes, Tristan; Sidus Space
Matthews, Jeff; Radian Aerospace
Mclaughlin, Scott; Spaceport America
Mello, Jason
Messer, Scott; United Launch Alliance
Mommer, Ric; DIU
Monteith, Wayne; Bechtel National, Inc.
Mosdell, Joy; Relativity Space
Olson, John , USSF
Papandrew, Devon; STOKE Space Technologies
Park, Helen; DIU
Patrick, Jason; Sidus Space
Poulos, Dennis; DIU
Ridenoure, Rex; DIU
Riley, Blaine; ABL Space Systems
Rokaw, Mike; Virgin Orbit National Systems
Rosenthal, Steven; California Military Department
Schenewerk, Caryn; Relativity Space
Shirah, Lisa; Aerodyne Industries, LLC
Skylus, Jay; Aevum, Inc.
Sleiman, Youssef; ENSCO
Smith, Shane; Operator Solutions
Walsh, Kevin; BRPH Mission Solutions
Wood, Nathan; Sierra Space

## VIRTUAL PARTICIPANTS
Albuquerque, NM

Baker, Julian; Atlantic Council
Battle, Joe
Berkson, Brad; RocketStar
Blanks, David; AFRL
Bruno, Michael; Aviation Week Network
Bullington, Joe; Jacobs Technology

Calnan, Gary; CisLunar Industries
Carlisle, Kirby; Argo Space Corp
Carol, Elliot; Lunar Resources, Inc.
Chandrabose, Vincent
Coble, Keith; Apogee Engineering
Cook, T.J.; Ingenuity Venture Fund
Cook, William; MDA
Davies, Iulia; Umbra
DeBonis, David; Atmospheric and Environmental Research
DeGeorge, Drew; AFRL Rocket Lab
DeHerrera, Arial; NSNM
DeVries, Bailey
Durr, Laura; USSF
Erwin, Sandra; SpaceNews
Estep, Nicholas; DIU
Fatemi, Navid; SolAero
Fletcher, Charles; NASA
Frost Adam; EXIM
Gabbert, Rob; Space PAC
Gaud, Summer
Gossel, Jeff; Jeff Gossel Consulting, LLC
Gregory, William; Southwest Creek Engineering
Halbach, Rick; lmco
Harrison, Soren; Fiore
Harvey, Brian; BA & associates
Hawley, Todd; Partners in Air and Space
Hecht, Erika; Market Ascent
Hehn, Trevor; Attorney
Hoff, Jamie; AFRL
Hoffman, Lars; Rocket Lab USA
Hooks, Daniel; LANL
Hudson, Jennifer
Jemison, Thomas; BNNano Inc
Karanian, Linda; Karanian Aerospace Consulting, LLC
Ketcham, Dale; Space Florida
Koleski, Katherine; DIU
Kron, Benjamin; York Space Systems
Kuruvilla, Saju; Northrop Grumman
Lal, Dr. Bhavya; NASA
Lasky, Robert; Astronet
Lawless, Juli; Redwire Space
Leszczynski, Zigmond; L10 Innovations
Lo, Eric; Booz Allen Hamilton
Loftus, Tom; Razor's Edge Ventures
M, Jason
Mahoney, Sean; Consultant
Marie, Aviation; AviationMarie LLC
Martin, Jeffrey; University of Alabama
Master, Todd; Umbra

Matheny, Jason
McClain, Sean; NRO/OP&S
McDonald, Kathleen; CloudArc
McIntyre, Jaime; Rescale
Meier, Steven; US NRL
Miller, Alex; DIU
Mital, Vivek; VegaMX
Moses, Robert; Axiom Space
Nagelin, Mike; Arcfield
Oelrich, Madeline; Redwire Space
Olson, John; USSF
Osborne, Kenneth; Space Systems Services, LLC
Park, Helen; DIU
Pillow, Katrina; NRO
Potter, Rick; AER
Rathje, Jason
Rollins, Rylee; Redwire Space
Shah, Jiral; Gravitics
Shinnick, Mathis; UNM Corporate Engagement
Shumaker, Nicole
Singh, Pavneet; DIU
Smith, Marlena
Stringer, Jeremy; MTRI
Thayer, Chris; Motiv Space Systems
Weeden, Brian; Secure World Foundation
Westerdahl, Ryan; Turion Space Corp
Williamson, Dave; Terran Orbital Corporation
Zingler, Scott; Tecolote Research

**VIRTUAL PARTICIPANTS**
Cape Canaveral, FL

Batis, Melissa; NASA
Bontrager, Mark; USSF
Brink, Jeff; NASA KSC
Burk, Kristin; SSC/S3
Carrillo, Ruben
Catledge, Burton
Chrisman, Tim; Foundation for the Future
DeHerrera, Arial; NewSpace NM
Drees, Karina; Commercial Spaceflight Federation
Duggleby, Andrew; US Navy / Venus Aero
Garcia, Celestino; ENSCO Inc
Gratias, Matthew; Relativity Space
Johnson, Darren; HQ SSC/S3
King, Jerry; SLD 45/XP
Kulin, Robb; STOKE Space Technologies
Light, Laura; SSC/S3
Marie, Aviation; AviationMarie LLC
Marotta, Tom; The Spaceport Company
McPhee, Mercedes; Commercial Spaceflight Federation

Mechtly, Victoria; RS and H
Mello, Jason; Firefly
Petro, Janet; NASA
Pierce, Jillianne; Space Florida
Posada, Anthony; Blue Origin
Rau, Steven; ManTech | SLD 30 Operations
Riddle, Randy; USSF
Shirah, Lisa; Aerodyne Industries LLC
Smith, Craig J; Oklahoma Air & Space Port
Tharpe, Jennifer; NASA
Vorbach, Ian; The Spaceport Company

# APPENDIX B
## PREVIOUS REPORTS

**State of the Space Industrial Base 2021**
Infrastructure & Services for Economic Growth & National Security

October 2021

Distribution A:
Approved for Public Release. Distribution Unlimited.

[Download]

**Defining the Road to 2035-45 USSF Capabilities**
Report on the USSF Space Futures Workshop 2a

5 Aug 2021

Distribution D:
Authorized to the Department of Defense and U.S. DoD contractors only

**State of the Space Industrial Base 2020**
A Time for Action to Sustain U.S. Economic & Military Leadership in Space

July 2020

Distribution A:
Approved for Public Release. Distribution Unlimited.

[Download]

**The Future of Space 2060 & Implications for U.S. Strategy**
Report on the Space Futures Workshop

5 Sep 2019

Distribution A:
Approved for Public Release. Distribution Unlimited.

[Download]

Download

**State of the Space Industrial Base: Threats, Challenges and Actions**
A Workshop to Address Challenges and Threats to the U.S. Space Industrial Base and Space Dominance

30 May 2019

Distribution A:
Approved for Public Release. Distribution Unlimited.

**Space Power Competition in 2060: Challenges and Opportunities**
Report on the Space Futures Workshop 1A

9 Mar 2019

Distribution D:
Authorized to the Department of Defense and U.S. DoD contractors only

# APPENDIX C
## KEY ACTIONS & RECOMMENDATIONS FROM SSIB'21 REVISITED

These recommendations reflected the numerous inputs from the 2021 workshop participants and their best assessment regarding which agency(cies) are in the best position to forward the necessary change.

1. **Establish "Space Development and Settlement" as our National "North Star" Space Vision** - The United States still requires a whole-of-nation vision and strategy for the economic and industrial development of space, to unite all elements of national power, and to attract like-minded allies and partners to a common wealth-creation framework.

   UPDATE: Though a North Star vision has not been issued, the Administration has released a Space Priorities Framework and ISAM Strategy.

2. **Build Back <u>Beyond</u>: Incorporate the Moon into the Earth's Economic Sphere by Catalyzing the Space Superhighway** - Building back better means an expanded economic canvas for America in the one theater that offers a million times Earth's resources and a billion-times its energy.

   UPDATE: Although NASA has provided initial plans for its Artemis Basecamp, and announced a launch date for Artemis 1, the Moon is still framed in terms of science and exploration goals rather than economic development goals. OSTP's recent request for information on developing Cislunar signals a potential change in paradigm. While a diversity of launch providers are working toward reusability (for example, Starship, New Glenn, Neutron and Relativity-R), FAA environmental delays have already compromised America's lead in reusable launch vehicles,[387] slowed the development of the Artemis Human Landing System, and put in jeopardy America's soft power.

3. **Sustain Funding for the Hybrid Space Architecture[388] as a Foundation for the Future Space Internet** - Congress provided critical funding in FY21 to seed a hybrid information network architecture in space capable of connecting disparate civil, military, commercial and allied systems in a secure environment that leverages the cloud, layered security, and low latency communications.

   UPDATE: While the broader meaning of the Hybrid Space Architecture as meant in this document is not adequately funded, a bright area is the funding for Tranche 0 and Tranche 1 of the Space Development Agency (SDA).

4. **Expand "Artemis Accords" Beyond NASA** - Establish international norms of behavior with Allies and partners to include operational space traffic management data exchanges, space domain awareness, and science and technology collaboration to promulgate democratic principles in space consistent with the rule of law, human rights, and liberties that underpin the norms, standards, and rules of acceptable behavior and action.

   UPDATE: The Administration has made significant moves to establish norms. More nations have signed the Artemis Accords, the U.S. is participating in the U.N. open-ended working group, and the

---

[387] Jones, A. (2022). China could shift to fully reusable super heavy-launcher in wake of Starship. SpaceNews.
[388] See Appendix F.

Vice President has announced a unilateral ban on kinetic anti-satellite missiles. The Secretary of Defense has released Tenets of Responsible Space Behavior. However, more needs to be done to establish international norms of behavior that will enable a free and open Cislunar region. More needs to be done to use U.S. Cislunar missions to set foundational precedents and standards. Space Policy Directive 1 directed NASA to: "Lead an innovative and sustainable program of exploration with commercial and international partners to enable human expansion across the solar system and to bring back to Earth new knowledge and opportunities. Beginning with missions beyond low-Earth orbit, the United States will lead the return of humans to the Moon for long-term exploration and utilization, followed by human missions to Mars and other destinations."[389] While a plethora of potential federal stakeholders have been identified,[390] an emphasis on commercial partnerships and utilization strongly suggests a larger role for the Department of Commerce. Many of the key partnerships on Space Traffic management and data exchanges require a stronger Department of Commerce.

5. **Increase Space Science & Technology Funding to Parity with Other Domains** - Neither our nation's spacefaring ambitions nor its economic or national security imperatives can be accomplished without technology leadership.

   UPDATE: Space S&T saw significant increases in both the enacted 2022 and proposed 2023 budgets.

6. **Reform Policy to Address 21st Century Conditions** - Most legacy policies were designed for a world with very different pressures and concerns. Technology and information ought to be a strength of the United States as a strategic partner. Yet legacy policies often result in the U.S. Government, its partners and allies being slow or late to receive the latest and best technology from the very dynamic and vibrant U.S. industrial base.

   UPDATE: The administration has significantly increased the funding for the Office of Space Commerce.

7. **Declare Space a Special Economic Zone and Deploy the Full Range of Tools** - It is time to declare space a special economic zone and to deploy the full range of financial tools and incentives to stimulate growth and industrial expansion.

   UPDATE: Space has not been declared a Special Economic Zone.

8. **Recognize Space-critical Infrastructure / Make Space a Part of Infrastructure Plans** - The fact that space systems constitute infrastructure critical to our way of life and prosperity is now broadly accepted, yet it is not formally recognized.

   UPDATE: Space has not been recognized as critical infrastructure.

9. **Make Space a Central Part of Climate Action Plans** - The space industrial base is capable of immense and diverse contributions to tackling climate change. But it must be deliberately mobilized by White House policy.

---

[389] White House (2017). Presidential Memorandum on Reinvigorating America's Human Space Exploration Program.
[390] Thomas D. Olszewski, T. & J. Locke (2020). Potential Roles of Federal Agencies in Creating a Sustainable Presence on the Moon. IDA.

UPDATE: Although the administration has resourced NASA climate observation and sought to apply remote sensing for monitoring, it has nothing comparable to the UK's Space Energy Initiative,[391] ESA's Solaris,[392] project or China's Space Solar Power program.[393]

10. **Include Space in Supply Chain Planning** - A failure to look ahead has put the U.S. behind on global infrastructure, 5G, rare-earth and medical supply chains. The U.S. space sector -- because it is the global leader -- is the subject of intense predation. As such, the Administration should prioritize space in its supply chain planning.

    UPDATE: President Biden signed the Creating Helpful Incentives to Produce Semiconductors (CHIPS) and Science Act on August 2, 2022. This legislation will aid in meeting supply chain shortages impacting the satellite industry.[394]

11. **Integrate JADC2 with the Hybrid Space Architecture** - It is critical that the DoD leverages the capabilities of the Hybrid Space Architecture to fully enable Joint All Domain Command and Control (JADC2).

    UPDATE: Two new activities began in earnest in 2021, an inflight cubesat experiment named Recurve led by AFR/RV,[395] and force design for the HSA led by the Space Warfare Analysis Center (SWAC).

12. **Enable the Space Superhighway by Including Commercial Solutions for In-space Logistics Infrastructure** - As USSF articulates its architecture for in-space mobility and logistics, it is critical to include commercial solutions from the start. This effort should be done in close collaboration with NASA.

    UPDATE: The USSF has yet to articulate a force design, architecture, concept of operations, or requirements for in-space mobility and logistics.

13. **Mandate a Percentage of Commercial Services Buys Starting in 2022** - Today commercial services procurements represent single digit percentages of the overall DoD acquisition budget.

    UPDATE: The reaction from industry to the FY23 budget was "I think it's fair to say that this budget doesn't reflect a pivot to a greater adoption of commercial capabilities in lieu of government-owned and operated capabilities. The giant pivot people were hoping for is just not happening, at least not as quickly as commercial operators would have liked."[396]

14. **Expand Use and Management of Space Commercial Services within the Space Force** - The unprecedented levels of capital investment in space innovation and small business creation will not be sustained if investors cannot see revenue or a return on investment.

---

[391] Space Energy Initiative (2022). Space Energy Initiative.
[392] ESA (2022). SOLARIS.
[393] Jones, A. (2022). China aims for space-based solar power test in LEO in 2028, GEO in 2030. SpaceNews.
[394] Rainbow, J. (2022). Biden to sign chips bill in a boost for satellite supply chains. SpaceNews.
[395] Dailey, J. (2022). AFRL spacecraft Recurve launches on Virgin Orbit Space Force mission. AFRL.
[396] Erwin, S. (2022). DoD Satcom: Big money for military satellites, slow shift to commercial services. SpaceNews.

UPDATE: The COMSO was set up with a broad mandate to purchase 'space as a service' for anything except launch.

15. **Bolder Acquisition Reform Means a More Level Playing Field for All Business, Particularly Small Business** - Most innovation, economic growth and jobs come from small business, however structural barriers "architect out" many would-be commercial providers.

    UPDATE: The USG must be cautioned. Venture Capital is less interested in supporting SBIR grantees than they are viable commercial startups with high revenue potential.

16. **Enable Rapid Innovation by Shifting Resources from SBIRs to OTAs** - While Small Business Innovative Research contracts (SBIRs) can help in the very early stages of a small business, delivering real capability requires a shift to mid & late term Other Transaction Authorities (OTAs) led by DIU and SpaceWERX, which are more effective instruments than Phase II or Phase III SBIRs for nurturing the national security innovation base.

    UPDATE: There has been no significant shift in priority from SBIR to OTA transactions. SBIR funding is a tax on RDT&E funding that would otherwise support OTAs at DIU and other organizations. DoD should be relieved of this tax burden in order to fund more prototypes, experiments and challenge activities through DIU and other innovation organizations.

17. **Balanced Growth Requires Investment Beyond LEO** - The innovations in venture capital and SPACs have put the Low Earth Orbit (LEO) economy on solid ground with a tremendous diversity of space access options and space information services. However, we do not yet have a MEO, HEO, GEO, Cislunar or Lunar economy that is new space-oriented.

    UPDATE: While there is increased investor interest beyond LEO, there remains an unbalanced investment in launch compared to broader space access and investment beyond GEO. Investors respond to demand signals from would-be customers. The USG needs to be more vocal about future needs and willingness to procure commercial solutions.

18. **Expand Investments in Enabling Technologies** - Key enabling technologies in the supply chain for the expanding space economy are missing or presently over-indexed off-shore. In particular, the supporting components such as microelectronics and scalable high-performance photovoltaics need to be developed and on-shored.

    UPDATE: The U.S. is still deficient in on-shore manufacturing of components to support the space supply chains. The CHIPS and Science Act helps, but more can be accomplished through meaningful procurement contracts focused on dual-use technologies.

# APPENDIX D
## SSIB'22 PARTICIPANTS SURVEY AND RESULTS

An electronic survey was performed by New Space New Mexico (NSNM) during the SSIB'22 Workshop. Participants were asked to subjectively rate observed progress on each of the SSIB'21 Key Actions and Recommendations using a scale of 1 to 10 to score progress. The average score of the 90 respondents is reflected in parentheses below.

| RECOMMENDATIONS | SCORE |
|---|---|
| **RECOMMENDATIONS FOR THE WHITE HOUSE & SPACE COUNCIL** | |
| 1. Establish "Economic Development and Human Settlement" as a National (bipartisan) "North Star" Space Vision for the 21st Century | 4.0 |
| 2. Build Back Beyond: Incorporate the Moon into the Earth's Economic Sphere by Catalyzing the Space Superhighway | 3.5 |
| 3. Sustain funding for the Hybrid Space Architecture as a foundation for the future Space Internet | 4.9 |
| 4. Expand "Artemis Accords" Beyond NASA | 4.6 |
| 5. Increase Space Science & Technology Funding to Parity with Other Domains | 4.3 |
| 6. Reform Policy to Address 21st Century Conditions | 3.5 |
| 7. Declare Space a Special Economic Zone and Deploy the Full Range of Tools | 2.8 |
| 8. Recognize Space-critical Infrastructure / Make Space a Part of Infrastructure Plans | 4.1 |
| 9. Make Space a Central Part of Climate Action Plans | 4.2 |
| 10. Include Space in Supply Chain Planning | 4.1 |
| **ATTENDEE RECOMMENDATIONS FOR THE DoD** | |
| 11. Integrate JADC2 with the Hybrid Space Architecture | 4.1 |

| | |
|---|---|
| 12. Enable the Space Superhighway by Including Commercial Solutions for In-space Logistics Infrastructure | 5.0 |
| 13. Mandate a Percentage of Commercial Services Buys Starting in 2022 | 4.1 |
| 14. Expand Use and Management of Space Commercial Services within the Space Force | 4.9 |
| 15. Bolder Acquisition Reform Means a More Level Playing Field for All Business, Particularly Small Business | 3.7 |
| 16. Enable Rapid Innovation by Shifting Resources from SBIRs to OTAs | 4.0 |
| **ATTENDEE RECOMMENDATIONS FOR VENTURE CAPITAL AND INVESTORS** | |
| 17. Balanced Growth Requires Investment Beyond LEO | 4.0 |
| 18. Expand Investments in Enabling Technologies | 4.6 |

## Attendee Perceptions of Progress

**Disappointment with Overall Progress** - Disappointingly, attendees were in general pessimistic about the progress made, subjectively rating progress as negative or at a stand-still for each recommendation.

Participants were moderately satisfied with the progress to enable the Space Superhighway by incorporating commercial solutions, the expanded use and management of space commercial services within Space Force, and sustaining funding for the hybrid structure.

They were moderately disappointed with progress to expand investment in enabling technologies, expanding "Artemis" beyond just NASA, increasing S&T funding to parity, making space a central part of climate action plans, integration with JADC2, including supply chain planning, mandating a percentage of commercial buys, and recognizing space as critical infrastructure.

*"It doesn't seem that the government has acted upon the key (SSIB'21) recommendations and that would have made things better. Good that there are some components, but the U.S. is still missing the North Star vision, incentives and finances."* -- SSIB'22 PARTICIPANT

Participants were greatly dissatisfied with the efforts to establish a space development and settlement vision, enable rapid innovation in SBIR/OTA, balance investments beyond LEO, reform policy to address 21st century problems, accomplish bolder acquisition reform, and incorporate the Moon into our economic sphere. Worst scoring of all was declaring space special economic zone and deploying the full range of tools.

## Attendee Perceptions of Priority

Participants were also asked to score last year's recommendations in terms of their importance and ease to accomplish.

**Most Important to Address** - The five most important by order were: Recognize Space-critical Infrastructure and make space part of infrastructure plans (43), Expand Investment in Enabling technologies (39), Establishing Space Development and Settlement as the National Vision (35), Incorporate the Moon into Earth's Economic Sphere by Catalyzing the Space Superhighway (33), Bolder Acquisition Reform (especially for small businesses) (32).

**Easiest to Accomplish** - The five recommendations judged easiest to accomplish in corder were: Expand investments in Enabling Technologies (39); Expand use and management of commercial services within the Space Force (35); Mandate a percentage of commercial services buys (35), Recognize Space-critical Infrastructure and make space part of infrastructure plans, (33), and Establish Space Development and Settlement as the National North Star vision (33).

**Important and Easy** - Those judged both important and easy included: Recognize Space as Critical Infrastructure (43+33=76), Incorporate Development and Settlement as the National Vision (35+33=68), and Expand Investments in Enabling Technology (39+39= 78).

*This page was intentionally left blank.*

# APPENDIX E
## ACRONYMS & ABBREVIATIONS

3D – Three Dimensional (printing)
3GPP – 3rd Generation Partnership Project
5G – Fifth Generation Wireless Internet
ADS-B – Automatic Dependent Surveillance–Broadcast
AFIT – Air Force Institute of TechnologyAFRL – Air Force Research Lab
AFRL/RV – Air Force Research Lab Space Vehicles Directorate
AFSS – Automated Flight Safety System
AFTS – Autonomous Flight Termination System
AIS – Automatic Identification System
ALINA – Autonomous Landing and Navigation Module [German PTScientists]
ALSEP – Apollo Lunar Surface Experiment Packages
ARPA-E – Advanced Research Projects Agency Energy [DoE]
ARPANET – Advanced Research Projects Agency Network
ASAT – Anti Satellite
ASCENT – Advanced Spacecraft Energetic Non-Toxic
AUKUS – Australia, the United Kingdom, and the United States (trilateral security pact)
BBC – British Broadcasting Corporation
BIS – Bureau of Industry and Security (DoC)
BMC3 – Battle Management, Command, Control, and Communications
BP – Bundle Protocol
CAPSTONE – Cislunar Autonomous Positioning System Technology Operations and Navigation Experiment [NASA]
CHIPS – Creating Helpful Incentives to Produce Semiconductors
CHPS – Cislunar Highway Patrol System [AFRL]
CLPS – Commercial Lunar Payload Services (NASA)
CNN – Cable News Network
CNSA – China National Space Administration
CO2 – Carbon Dioxide (atmospheric gas)
COMSO – Commercial Services Office (under USSF/SSC)
CONFERS – Consortium for Execution of Rendezvous and Servicing Operations
COVID-19 – Coronavirus Disease 2019
CSCO – Commercial Satellite Communications Office (USSF)
CSfC – Commercial Solutions for Classified (cryptography)
CSIS – Center for Strategic and International Studies (think tank)
CSO – Commercial Solutions Openings
CSO – Chief of Space Operations [USSF]
CSP – Communication Services Program (NASA)
CSPO – Commercial Systems Program Office (NRO)
CTEP – China & Transformational Exports Program (CTEP) [ExIm]
CTO – Chief Technology Officer
CYBERCOM – Cyber Command
DAF – Department of the Air Force
DARPA – Defense Advanced Research Projects Agency [DoD]
DDTC – Directorate of Defense Trade Controls [DoS]
DFARS – Defense Federal Acquisition Regulation Supplement
DFC – U.S. International Development Finance Corporation
DHS – Department of Homeland Security
DIU – Defense Innovation Unit [DoD]
DoC – Department of Commerce
DoD – Department of Defense
DoE – Department of Energy
DoS – Department of State
DoT – Department of Transportation

DPA – Defense Production Act
DPAS – Defense Priorities and Allocation System
DPC – Domestic Policy Council [EOP]
DRACO – Demonstration Rocket for Agile Cislunar Operations (DARPA)
DTN – Delay/Disruption Tolerant Networking
DTSA – Defense Technology Security Administration [DoD]
ELSA-d – End-of-Life Services by Astroscale-demonstration (Astroscale)
EO – Electro Optical
EO/IR – Electro Optical / Infrared (camera)
EOP – Executive Office of the President
EPA – Environmental Protection Agency
EU – European Union
ExIm – Export Import Bank
FAA – Federal Aviation Administration [DoT]
FBI – Federal Bureau of Investigation
FCC – Federal Communication Commission
FDOT – Florida Department of Transportation
FEMA – Federal Emergency Management Agency
FOCI – Foreign Ownership, Control or Influence
FTE – Full-Time Equivalent
FY – Fiscal Year
GAO – General Accounting Office
GDP – Gross Domestic Product
GEO – Geostationary Earth Orbit
GEOINT – Geospatial Intelligence
GPS – Global Positioning System
HSA – Hybrid Space Architecture
IAC – International Astronautical Congress
IAWG – Interagency Working Group
IC – Intelligence Community
ICAO – International Civil Aviation Organization
ILRS – International Lunar Research Station
IP – Intellectual Property
IPO – Initial Public Offering
ISAM – In-Space Servicing Assembly and Manufacturing
ISRU – In-Situ Resource Utilization
ISS – International Space Station

iSSI – iBoss Intelligent Space Systems Interface
ISWG – Interagency Spaceport Working Group
ITAR – International Trafficking in Arms Regulation
JADC2 – Joint All Domain Command and Control
JWST – James Webb Space Telescope (NASA)
kWe – Kilowatt-electric
LEO – Low Earth Orbit
LNG – Liquid Natural Gas
LOX – Liquid Oxygen
MDA – Missile Defense Agency
MEO – Middle Earth Orbit
MEV – Mission Extension Vehicle
MMPACT – Moon-to-Mars Planetary Autonomous Construction Technology
MOSAR – MOdular Spacecraft Assembly and Reconfiguration [EU]
MRV – Mission Robotic Vehicle
MW – Megawatt
NAS – National Air Space
NASA – National Aeronautics and Space Agency
NATO – North Atlantic Treaty Organization
NC3 – Nuclear Command, Control & Communications
NDAA – National Defense Authorization Act
NDSA – National Defense Space Architecture
NEC – National Economic Council [EOP]
NESDIS – National Environmental Satellite Data and Information Service
NGA – National Geospatial Agency [DoD]
NIST – National Institute of Standards and Technology
NOAA – National Oceanic and Atmospheric Agency [DOC]
NOTAMs – Notices to Airmen
NOTMARS – Notices to Mariners
NRHO – Near-Rectilinear Halo Orbit
NRL – Naval Research Lab [DoD]
NRO – National Reconnaissance Organization [DoD]
NSA – National Security Agency
NSAM – National Security Action Memorandum

NSC – National Security Council [EOP]
NSF – National Science Foundation
NSIC – National Security Investment Capital [DoD]
NSNM – NewSpace New Mexico
NSpC – National Space Council [EOP]
NSS – National Space Society
NTP – Nuclear Thermal Propulsion
O&M – Operations & Maintenance
OCEA – Office of Commercial and Economic Analysis (Air Force)
ODNI – Office of the Director of National Intelligence
OISL – Optical Intersatellite Link
OMB – Office of Management and Budget [EOP]
OPM – Office of Personnel Management
OPR – Office of Primary Responsibility
OSAM – On-Orbit Servicing Assembly and Manufacturing
OSC – Office of Commercial Space [DoC]
OSD – Office of the Secretary of Defense [DoD]
OSTP – Office of Science and Technology Policy [EOP]
OTA – Other Transaction Authority
OUSD R&E – Office of the Undersecretary of Defense for Research and Engineering
PDM – Program Decision Memorandum
PEA – Programmatic Environmental Assessment
PhD – Doctorate of Philosophy
PIC – Program Integration Council
PPBS – Planning, Programming, and Budgeting System
PPF – Payload Processing Facility
PRC – People's Republic of China
PSSI – Prague Security Studies Institute
PULSAR – Prototype for an Ultra Large Structure Assembly Robot [EU]
PV – Photovoltaics
QD – Quantity Distance
QR – Quick Response (a square bar-code)
R&D – Research and Development
RAFTI – Rapidly Attachable Fluid Transfer Interface

RAPIDS – Robust Access to Propellant in Diverse orbitS [DIU]
RDT&E – Research Development Test and Evaluation
RF – Radio Frequency
ROM – Rough Order of Magnitude
ROS – Robot Operating System
ROTF – Range of the Future
RTG – Radioisotope Thermoelectric Generators
S&E – Science and Engineering
S&T – Science and Technology
SAF/IA – Secretary of the Air Force International Affairs
SAF/SQ – Office of the Assistant Secretary for Space Acquisition and Integration
SAML – Space Access Mobility and Logistics
SAR – Synthetic Aperture Radar
SATCOM – Satellite Communications
SBA – Small Business Administration
SBIC – Small Business Investment Companies
SBIR – Small Business Innovative Research
SBSP – Space Based Solar Power (renewable energy source)
SCO – Strategic Capabilities Office [DoD]
SCP – Space Capstone Publication
SDA – Space Development Agency
SDA – Space Domain Awareness
SECDEF – Secretary of Defense
SLIM – Smart Lander for Investigating the Moon [JAXA]
SML – Space Mobility and Logistics
SNAP-27 – System for Nuclear Auxiliary Power
SPAC – Special Purpose Acquisition Corporation
SpaceX – Space Exploration Technologies (company)
SpEC – Space Enterprise Consortium
SpRCO – Space Rapid Capabilities Office
SSA – Space Situational Awareness
SSC – Space Systems Command [USSF]
SSP – Space Solar Power
SSIB – State of the Space Industrial Base (report)
SSN – Space Surveillance Network

SSPIDR – Space Solar Power Incremental Demonstrations and Research Project [AFRL]
STEAM – Science, Technology, Engineering, Arts and Mathematics
STEM – Science Technology Engineering and Math
STM – Space Traffic Management
STRATFI – Strategi Funding Increase [USAF SBIR program]
STTR – Small Business Technology Transfer (STTR)
SWAC – Space Warfighting Analysis Center
TACFI – Tactical Funding Increase [USAF SBIR program]
TCAS – Traffic Collision Avoidance System
TCP/IP – Transmission Control Protocol/Internet Protocol
TCPED – tasking, collection, processing, exploitation, and dissemination tasking, collection, processing, exploitation and dissemination
TNT – Trinitrotoluene (explosive)
TOA – Total Obligation Authority (budget)
TRL – Technology Readiness Level
TS/SCI – Top Secret / Sensitive Compartmented Information
TTC – Trade and Technology Council
U.S. – United States
UK – United Kingdom
UKN – Unknown
USA – United States of America
USAF – United States Air Force
USCG – United States Coast Guard
USD(A&S) – Under Secretary Of Defense For Acquisition And. Sustainment
USD(R&E) – Office of the Under Secretary of Defense, Research and Engineering
USDOT – United States Department of Transportation
USERC – University Space Engineering Research Centers
USG – United States Government
USGS – United States Geological Service
USSF – United States Space Force [DoD]
USSPACECOM – United States Space Command [DoD]

VIPER – Volatiles Investigating Polar Exploration Rover [NASA]
Yr – Year

*This page was intentionally left blank.*

www.ingramcontent.com/pod-product-compliance
Lightning Source LLC
Chambersburg PA
CBHW082123230426
43671CB00015B/2790